HEALTH &
SAFETY IN
PRACTICE

Human Factors and Safety

Jeremy Stranks

PITMAN
PUBLISHING

PITMAN PUBLISHING
128 Long Acre, London WC2E 9AN

A Division of Pearson Professional Limited

First published in Great Britain, 1994

© Jeremy Stranks, 1994

British Library Cataloguing in Publication Data
A CIP catalogue record for this book can be obtained from the British Library.

ISBN 0 273 60440 6

10 9 8 7 6 5 4 3 2

Typeset by Northern Phototypesetting Co Ltd., Bolton
Printed and bound in Great Britain by
Bell & Bain Ltd., Glasgow

The Publishers' policy is to use paper manufactured from sustainable forests

Contents

Preface

Traditionally, health and safety legislation, such as the Factories Act 1961 and the Offices, Shops and Railway Premises Act 1963, has concentrated on the provision and maintenance by an employer of a safe place of work. This has been accomplished by the continuing enforcement of legal requirements relating to machinery and work equipment, construction operations, the use of hazardous substances, environmental factors and a host of other 'safe place' requirements. Very little attention was paid in the past, however, to people in terms of their individual needs at work, attitudes to health and safety in the work situation, their instruction and training, individual capabilities and the chance of their making mistakes, and the role of people generally in the accident causation process. The Health and Safety at Work Act 1974 (HSWA) went some way towards redressing this situation in that employers are required to provide safe systems of work, information, instruction and training, but, in retrospect, not far enough.

The Management of Health and Safety at Work regulations 1992 (MHSWR), for the first time in health and safety legislation, bring in a human factors-related approach to occupational health and safety, along with the general duty of employers to actually manage health and safety at work.

While regulation 4 of the MHSWR places this general duty at the feet of employers to actually manage their health and safety activities through 'effective planning, organization, control, monitoring and review of the preventive and protective measures', regulation 11 deals with 'capabilities and training'. The situation relating to 'capabilities' is, it quite simply states:

Every employer shall, in entrusting tasks to his employees, take into account their capabilities as regards health and safety.

What, though is a 'capable' person. Various terms are found in the average dictionary – 'able', 'competent', 'gifted' and 'having the capacity'. Perhaps this last term is the most significant from a health and safety viewpoint. 'Capacity' implies both mental and physical capacity – mental capacity to understand why a task should be undertaken in a particular way and physical capacity, the actual strength and fitness to undertake the task in question. The need for this human factors-related approach must, therefore, be considered in a whole range of situations, such as in the design of safe systems of work and work equipment, the organization of the workplace, the control of the working environment and in planning the organizational structure for occupational safety and health generally.

Jeremy Stranks, 1994

List of abbreviations

ACOP	Approved Code of Practice
COSHH	Control of Substances Hazardous to Health Regulations 1988
DSE	Display screen equipment
HRA	Human reliability assessment
HSE	Health and Safety Executive
HSWA	Health and Safety at Work Act 1974
ILCI	International Loss Control Institute
LCU	Life change unit
MHSWR	Management of Health and Safety at Work Regulations 1992

What are human factors?

INTRODUCTION

'Human factors' are defined in the HSE publication 'Human factors in industrial safety' (HS(G)48) as covering a wide range of issues, which include:

- the perceptual, physical and mental capabilities of people and the interaction of individuals with their job and working environments

- the influence of equipment and system design on human performance and, above all,

- the organizational characteristics that influence safety-related behaviour at work.

For all levels of management to take these issues into account, organizations need to establish and develop an appropriate 'safety culture' (see Chapter 10), which, for instance, discourages the taking of risks by individuals or groups, examines, and endeavours to correct, situations where the potential for human error is significant and ensures the provision of adequate health and safety information and training to operators and managers alike.

Thus, human factors have a direct relationship with the behavioural sciences, the range of sciences mostly concerned with the study of human behaviour at different levels or in specific ways, for example, psychology, social psychology, politics, ergonomics, economics, anthropology, sociology and psychiatry.

HUMAN FACTORS AND ACCIDENT PREVENTION

The pre-accident situation

In considering the relationship of human factors to accident prevention, it is important to identify the basic features of the typical pre-accident situation. These are:

- the *objective danger* at this point in time, that is, the shortcomings or deficiencies in the physical conditions – the badly fenced machine, slippery floor, unfenced floor opening

- the *subjective perception of risk* on the part of the individual or potential accident victim.

It is now widely accepted that the majority of accidents are, in some measure, attributable to human as well as technical factors, in the sense that actions by people initiate or contribute to the accidents, or people might have acted better to avert them.

The maintenance of a *safe place* strategy is readily identified in both former and current health and safety legislation, and the role of the enforcement authorities has always been geared to promoting improvements in physical conditions in the workplace. A system that relies heavily on the enforcement of minimum legal standards to bring about improvements in these physical conditions, however, has not always been successful in reducing accidents or ill-health due to this failure to examine the 'people' side of safety and accident prevention. Human factors, therefore, are concerned with *safe person* strategies, with the principal objective of increasing people's perception of the risks at work in order to prevent accidents.

The role of management

Although most UK health and safety legislation places the duty of compliance firmly on the body corporate – the organization – this duty can only be discharged by the effective actions of its managers. Studies by the HSE's Accident Prevention Advisory Unit have shown that the vast majority of fatal accidents, and those resulting in major injuries, such as fractures or amputations, could have been prevented by management action. For example:

- during the period 1981–5, 600 people were killed in the construction industry: 90 per cent of these deaths could have been prevented and in 70 per cent of cases, positive action by management could have saved lives
- a study of 326 fatal accidents during maintenance activities occurring between 1980 and 1982 showed that, in 70 per cent of cases, positive management action could have saved lives
- a study of maintenance activities in the construction industry between 1982 and 1985 demonstrated that 75 per cent were the result of management failing to take reasonable steps.

These studies tend to emphasize the crucial role of the organization in the management of job and personal factors and are principally aimed at preventing human error, but how many organizations actually consider the potential for human error in the design of jobs? The only indicator may be a high level of rejects in the finished product or some form of customer dissatisfaction. Clearly, management at all levels must recognize the need to motivate, train, supervise, inform and instruct workers in sound safety procedures with a view to reducing accidents at work. The benefits, in terms of improved performance and profitability, must be the greatest motivators for such action.

Influences on people at work

There are three principal influences on people at work:

- the organization
- the actual jobs that people do
- personal factors.

These areas are directly affected by:

- the system for communication within the organization
- the training systems and procedures in operation

all of which are directed at preventing human error.

2

The organization

FEATURES OF ORGANIZATION

One definition of the 'organization' is 'an arrangement of human and physical resources for the purposes of carrying out transactions with the environment in such a manner that it ensures the survival of that organization'. The principal objectives are the successful completion of tasks and survival.

Organizations take many forms. However, most organizations operate on the basis of a hierarchy, a set of relationships between groups of people. Most organisations operate on a pyramidal hierarchy, which is one based on power, authority and clearly established goals or objectives. Thus, orders pass down the system and information passes back up the system. The flow is one way in both cases. Moreover, promotion within the hierarchy is allegedly based on merit and hard work.

Organizations are, in many cases, a series of integrated systems covering the various inputs (knowledge, skills, raw materials, capital and legal requirements), management control systems, decision-making processes, general and specific actions, and the outputs of the organizations, which may include the supply of goods and services. Such systems work within a specific environment, which may be affected by physical, social, market and economic factors. They generally incorporate either a formal or informal structure.

A *formal organization* incorporates a system of positions, roles and interrelated groups that is designed as the most efficient arrangement for effectively accomplishing the aims of the organization.

An *informal organization*, on the other hand, incorporates the patterns of interpersonal and intergroup relationships that develop within a formal organization. This tends to develop when the formal organization proves to be ineffective or fails to satisfy important individual or group needs.

A typical formal organization incorporates a number of features.

- it is deliberately impersonal
- it is based on ideal relationships
- It may be based on the 'rabble hypothesis' of the nature of man, that is, 'each man for himself'.

It comprises, as a rule, the following.

1 *The functional/line organization* This is based on the type of work being done, such as manufacture of a product at its various stages, provision of a specific service.
2 *The staff organization* This is in direct contrast to the line organization. It is represented by people who have advisory, service and control functions, such as health and safety specialists, transport managers and accountants, respectively.

Potential sources of conflict

Conflict may arise between the line and staff organizations for a number of reasons, such as:

- differing motivations, such as the classic conflict between preventive maintenance engineers, whose principal objective is to keep the plant running, and health and safety specialists, who may be more concerned with safe working procedures
- misunderstanding of individual roles, whereby people may have different perceptions and notions about a particular person's role, based, perhaps, on a misunderstanding or lack of knowledge of what is involved
- differing cultures and objectives in that, for instance, the management group culture may be totally at odds with the shop floor and trade union-based cultures
- differing priorities and levels of commitment brought about by variations in the extent of individual, work group and staff motivation.

Weaknesses of formal organizations

The formal organization is by no means perfect. Typical weaknesses include communications failures. Moreover, they ignore emotional factors in human behaviour and are frequently seen by the workforce as uncaring and lacking in interest or commitment to ensuring appropriate levels of health, safety and welfare within the organization. These factors alone have a direct influence on the way people behave at work, particularly when confronted with hazards. It is important, therefore, to consider the organizational characteristics that actually influence safety-related behaviour.

Characteristics of organizations that influence safety-related behaviour

Studies throughout industry and commerce have shown that there are a number of factors, features or characteristics of organizations that influence the way people behave at work, particularly with regard to the operation of safe working procedures.

There is no doubt that the organisation (company, local authority, partnership) has a direct effect on human behaviour at work. In particular, it endeavours to control both individual and group behaviour in many ways. This may be through systems for supervision and control, training activities, management development systems, apprentice schemes, the reward structure of the organization and rates of pay, including bonuses and overtime payments.

The organizations that are most successful in their health and safety operations, and in reducing accidents, ill-health at work and the ensuing sickness absence that follows, incorporate the following features or characteristics:

- the promotion of a positive **climate** in which health and safety are seen by both management and employees as being fundamental to the organization's day-to-day operations, that is the creation of a positive *safety culture*

- ensuring that *policies and systems*, which are devised for the control of risk from the organization's operations, take proper account of human capabilities and fallibilities

- *commitment* to the achievement of progressively higher standards, which are shown at the top of the organization and cascaded down through successive levels

- *demonstration* by senior management of their active involvement, thereby galvanizing managers throughout the organization into action
- *leadership*, whereby an environment is created that encourages safe behaviour.

There is a need, therefore, for organizations to consider the above characteristics and establish systems that will indicate or predict potential human failure situations in work situations. This can best be achieved through safety monitoring systems (safety audits and inspections), joint consultation with staff, the assessment of risks, the design and implementation of safe systems of work, training and designation of competent persons for certain tasks and ensuring appropriate and meaningful feedback from the investigation of accidents and loss-producing incidents. The benefits in terms of better industrial relations, reduced accident and absence costs, increased efficiency and profitability, together with an improved marketplace image, should not be overlooked.

2

The appropriate safety climate

The HSE publication 'Human factors in industrial safety' states the need on the part of organizations to create a climate that promotes staff commitment to health and safety and emphasizes that deviation from corporate safety goals, at whatever level, is not acceptable (assuming that it has set corporate goals and these are correct in relation to safety).

Producing such a climate requires clear, visible management commitment to safety from the most senior levels in the organization. This commitment should be not just a formal statement, but be evident in the day-to-day activities of the enterprise, so that it is readily known and understood by employees. Individuals may be reluctant to err on the side of caution in matters that have safety implications if their decision to do so is likely to be subject to unwarranted criticism from their superiors or their peers.

The attitude of a strong personality at senior management level within the organization may have either a beneficial or an adverse effect on the safety climate. Inevitably, junior employees will be influenced by this person's example.

Safety procedures soon fall into disuse if there is no system for ensuring that they are followed. Too often, procedures lapse because of management neglect or operators are discouraged from working to them by peer group or other pressures, such as production targets. Where managers become aware of deficiencies in safety procedures, but take no action to remedy them, the workforce readily perceive that such actions are condoned.

Individuals may not understand the relevance of procedures or appreciate their significance in controlling risk. Sometimes procedures are faulty, irrelevant or lack credibility. When accidents happen, managers cannot blame individuals for taking short cuts that seemed safe and were allowed to become routine if they have not explained the importance of or monitored procedures they originally laid down.

To promote a proper working climate, it is essential to have an effective system for monitoring safety that identifies, investigates and corrects deviations. This could be through the operation of safety audits, safety inspections or surveys. There should be clearly defined standards and goals that are capable of being monitored, and effective systems for reporting and investigating incidents, near misses and unsafe situations.

The introduction and operation of such systems requires considerable effort by managers and only by allocating adequate resources can they be confident that failures will be prevented or controlled. In short, the organization needs to provide:

- clear and evident commitment from the most senior management downwards, which promotes a climate for safety in which management's objectives and the need for appropriate standards are communicated and in which constructive exchange of information at all levels is positively encouraged

- an analytical and imaginative approach, identifying possible routes to human factors failure (this may well require access to specialist advice)

- procedures and standards for all aspects of critical work and mechanisms for reviewing them

- effective monitoring systems to check the implementation of the procedures and standards
- incident investigations and the effective use of information drawn from such investigations
- adequate and effective supervision with the power to remedy deficiencies when found.

Management involvement

Management commitment is, perhaps, the most important of the points listed above. This commitment can be demonstrated as part of the process of encouraging a positive safety culture by:

- the Board clearly stating its intention, expectations and beliefs in relation to health and safety
- appropriate resources being made available in order to translate plans into positive achievements
- managers being accountable for their performance so far as health and safety is concerned and senior managers must be seen to take an active interest in the whole health and safety programme
- the reward of positive achievements
- actively involving lower levels of management who must, in turn, ensure that health and safety has a high profile within their areas of responsibility.

Getting results

Clearly, if management has little or no intention of becoming involved in health and safety, then any expressed commitment to it is unlikely to be taken seriously, and employees will not be motivated to use safe working procedures.

It is appropriate at this stage to state the '10 Principles of Safety' of the Du Pont Corporation, a world leader in the field of health and safety at work.

1 All injuries and occupational illnesses can be prevented.
2 Management is directly responsible for preventing injuries and illness, with each level accountable to the one above and responsible for the level below.

3 Safety is a condition of employment, that is, each employee must assume responsibility for working safely. Safety is as important as production, quality and cost control.

4 Training is an essential element for safe workplaces. Safety awareness does not come naturally – management must teach, motivate and sustain employees' safety knowledge to eliminate injuries.

5 Safety audits must be conducted. Management must audit performance in the workplace.

6 All deficiencies must be corrected promptly, either by modifying facilities, changing procedures, providing better employee training or disciplining constructively and consistently. Follow-up audits are to be used to verify effectiveness.

7 It is essential to investigate all unsafe practices and incidents with injury potential, as well as injuries.

8 Safety off the job is as important as safety on the job.

9 It is good business to prevent illnesses and injuries. They involve tremendous cost–direct and indirect. The highest cost is human suffering.

10 People are the most critical element in the success of a safety and health programme. Management responsibility must be complemented by employees' suggestions and their active involvement.

Many organizations would do well to emulate the example given by Du Pont and their Chief Executive, who personally undertakes safety audits.

LEADERSHIP AS A FEATURE OF ORGANIZATIONS

The theories of the late Douglas McGregor

In his book *The Human Side of Enterprise*, Douglas McGregor drew attention to a number of factors that are significant as far as organizations are concerned.

First, human beings know how to use physical science and technology to benefit themselves. The major problem now is how to use the social sciences to make human organizations effective. Given proper conditions, vast resources of creative energy can be made available.

Theory X

The conventional understanding of management's task is that of harnessing human energy to organizational requirements. This is based on the belief that without active intervention by management, people would be passive, even resistant, to organizational needs and that they are, by nature, indolent, lacking in ambition, self-centred and resistant to change. It follows from this belief that to achieve the best results with people, management must direct efforts and motivate and control the actions of subordinates in order to modify their behaviour to fit the needs of the organization. To achieve this objective, management's actions must fall between two extremes – the Hard Approach and the Soft Approach.

The *Hard Approach* is 'strong management', that is, directing the individual by coercion and threats, close supervision and tight controls. The problem with it is that 'force breeds counterforce', that is tough management creates antagonism and disinterest on the part of workers. In times of full employment, it also creates a high rate of labour turnover.

The *Soft Approach* is 'weak management', a permissive approach aimed at achieving a tractable labour force prepared to accept management's directions. In practice, people take advantage of this approach, taking more and giving less.

McGregor believed that Theory X was based on a correct observation of human behaviour, but that it is incorrect to believe that this behaviour is innate. Rather, these attitudes to work have been brought about by people's experience of the conditions that exist in organizations, in which case they can be modified by changing the conditions.

Theory X can be summarized as follows:

- the average human being has an inherent dislike of work and will avoid it if they can
- because of this human characteristic of dislike of work, most people must be coerced, controlled, directed and treated with punishment to get them to put forth adequate effort towards the achievement of organisational objectives
- the average human being prefers to be directed, wishes to avoid responsibility, has little ambition and wants security above all.

Managers who subscribe to Theory X will place great emphasis on pay and job security. They will not understand, for instance, that most of their employees are really trying to satisfy other motives, such as those relating to self-esteem and the realization of potential, or self-fulfilment (see Theories of motivation, Chapter 4). Their policies will be directed towards the manipulation of subordinates through the use of various incentives – money, in particular – that are largely ineffective in improving worker performance and subsequent production figures.

Theory Y

Modern organizations provide good pay, a fair level of security, good working conditions and adequate welfare facilities. The problem is that management does not get the cooperation and high levels of productivity that it expects because higher needs are not satisfied and employees experience frustration as a result.

Theory Y is McGregor's own theory, based on new assumptions about human motivation and behaviour. It is based on the belief that people are not passive or resistant to organizational needs by nature, but have been made so by their experience of working for organizations. The motivation, the capacity for assuming responsibility and directing behaviour towards organizational goals, are all present in people; management does not put them there.

It follows, then, that management's task is to make it possible for people to recognize and develop these characteristics for themselves. The essential task in this is to arrange organizational conditions and methods of operation so that people can best achieve their own goals, that is, the satisfaction of their needs, by directing their own efforts towards organizational objectives.

Certain concepts in accord with Theory Y are:

- *decentralization and delegation* this removes people from overly close control and satisfies their egoistic needs by giving them responsibility

- *job enlargement* this provides greater responsibility and status at the bottom of the organization

- *participation and consultive management* this involves people in decisions and group situations and makes it possible for management to tap the knowledge of subordinates; it also satisfies the subordinates' social and egoistic needs

- *performance appraisal* the Management Performance Standards Approach involves the agreeing of 'targets', or objectives, with the individual so that self-evaluation is possible.

Theory Y can be summarized thus:

- the expenditure of physical and mental effort in work is as natural as play or rest
- external control and the threat of punishment are not the only means for bringing about effort towards organizational objectives as people will exercise self-direction and self-control in the service of objectives to which they are committed
- commitment to objectives is a function of the rewards associated with their achievement
- the average human being learns, under proper conditions, not only to accept but to seek responsibility
- the capacity to exercise a relatively high degree of imagination, ingenuity and creativity in the solution of organizational problems is widely, not narrowly, distributed in the population
- under the conditions of modern industrial life, the intellectual potentialities of the average human being are only partially utilized.

The manager who works to Theory Y attempts to unite the goals of the individual and the organization, so that the employee does not need to be coerced into work. This does not imply that the Theory Y approach is a 'soft' approach, however; it is one based on human motivation.

Classical organization theory

Classical organization design is based on Theory X. It can be summarized in eight precepts (Bass, 1965):

1 some one person should be responsible for each essential activity
2 responsibility for each activity should not be duplicated and should not overlap
3 each position should have a limited number of clearly stated duties
4 every person should know exactly what their duties are

5 authority for making decisions should be commensurate with responsibility for them

6 authority should be delegated to persons close to the point of action

7 managers should have a limited number of subordinates, say, four to seven

8 every manager should know to whom they report and who reports to them – the chain of command should be clearly defined.

While the above concepts may be appropriate in a number of cases generally and specifically in the field of health and safety, the classical organization theory has a number of deficiencies. For instance, it fails to develop each individual's potential. Essential activities can be neglected at times, leading to major effort or even disasters. Furthermore, it, typically, creates a vertical structure, with the resulting communications problems.

On the other hand, this theory may be appropriate in certain areas of health and safety. For instance, in the identification of individual responsibilities, as outlined in a company statement of health and safety policy.

Characteristics of highly productive managers and organizations

The following characteristics are classed as important for such managers and organizations.

- a preponderance of favourable attitudes on the part of each member towards other members, superiors, the work and the organization as a whole

- identification with the objectives and a sense of involvement in achieving them

- all motivational forces:
- the ego, that is, the desire to achieve and maintain a sense of personal worth and importance
- desire for security
- curiosity, creativity and the desire for new experiences
- economic factors are harnessed.

- The organization consists of a tightly knit, effectively functioning social system comprised of interlocking work groups, and high levels of skill in personal interaction permit effective participation in group decision making

● measurements of organizational performance are used primarily for self-guidance rather than to superimpose control.

Management style

The answer to the question which style of management is more effective depends whether or not certain assumptions concerning managerial behaviour can be confirmed. For example:

● employee-centred supervisors are more productive than job-centred supervisors

● supervisors with high goals and a contagious enthusiasm as to the importance of achieving these goals achieve the best results

● there is a marked inverse relationship between the average amount of 'unreasonable' pressure the employees in the department feel and the productivity of that department

● general, rather than close, supervision is conducive to high productivity

● highly productive managers make clear to their subordinates what the objectives are and what needs to be accomplished, then give them the freedom to get on with the job

● the highly productive foreperson has a helpful, non-punitive attitude to errors

● there is a clear correlation between productivity and the workers' perception of the supervisor rather than between productivity and the general attitude towards the company

● high productivity is associated with skill in using group methods of supervision

● work groups can have goals that will influence productivity and cost either favourably or otherwise

● there is a clear relationship between group solidarity and productivity

● to function effectively, a supervisor must have sufficient influence with their own superior to be able to affect the superior's decisions.

Management style is, therefore, an important feature of an organization. While styles vary considerably, many of the above factors are significant to the success of the organization as a whole.

The autocrat and the democrat

In order to investigate an aspect of group functioning under different types of leadership and different types of group 'atmosphere', experiments with groups of children were undertaken in the 1930s by R. Lippitt and R. K. White. The aim was to establish small groups of children led by adult experimenters who adopted two totally different leadership styles, that is 'authoritarian' and 'democratic'. The groups were involved in various tasks and meetings were held regularly over a period of weeks.

The two different approaches incorporated the following features:

- *authoritarian style*:
  - all policies were determined by the appointed leader
  - techniques and steps for attaining the goals were dictated by the leader, one at a time; future direction was uncertain to a large degree
  - the leader usually dictated the work task and work companions of each member
  - the leader was 'personal' in praise and criticism without giving objective reasons
  - the leader remained aloof from active group participation except when demonstrating

- *democratic style*:
  - policies were determined as a result of group decisions
  - an explanation of the overall process was given at the first meeting
  - when technical advice was needed, the leader suggested several alternatives from which choices could be made
  - members were free to choose work companions and the division of tasks
  - the leader was 'objective' or 'fact-minded' in praise and criticism and tried to be a regular group member in spirit without doing much of the actual work.

The outcome of this experiment indicated that differences in the behaviour of the authoritarian-driven group and democratically led group were significant. The following are some of the more interesting differences.

- The authoritarian group tended to be either more aggressive or more apathetic than the democratic group. When aggression was expressed, it tended to be directed at other group members rather than towards the leader. Two scapegoats were the targets of such concentrated hostility that they left the club. In the apathetic authoritarian group, it seemed that the lack of aggression was due merely to the repressive influence of the leader as when the leader left the group for a short period, aggressive outbursts occurred

- In the authoritarian groups, there were more submissive approaches to the leader and more attention-seeking approaches. The approaches to the democratic leader were more friendly and task-related

- In the authoritarian groups, the relations between group members tended to be more aggressive and domineering than in the democratic groups

- Group unity appeared higher in the democratic groups and subgroups tended to be more stable than in the authoritarian atmosphere, where they tended to disintegrate

- Constructiveness of work decreased sharply when the authoritarian leader was temporarily absent, whereas it dropped only slightly when the democratic leader was absent

- When frustrations were induced in the work situation, the democratic groups responded by making organized attacks on the difficulty, whereas the authoritarian group tended to be disrupted by the recrimination and personal blame that occurred as a result.

BRINGING ABOUT CHANGE

Introducing change within the organization generally or for specific groups or individuals is a standard requirement if organizations are to survive. Generally, people are averse to change in any form, such as a change in job responsibilities, work location or work activities.

Organizational change

All organizations change from time to time, resulting in revised management structures, processes and procedures. Change, for most people, is a stressful experience – the exact level of stress caused varying from person to person. Organizational change should, therefore:

- allow for a period of adaptation by those affected by the change
- involve training of staff in new responsibilities
- be accompanied by continuing communication at the various stages of the change process
- be accepted by individuals and groups in terms of the actual need for change

if the process is to be successful. In most cases, the process involves some form of change management,which takes into account not only the financial implications of organizational change, but also the effects on performance that can result from such change.

Systems change

Prior to any change in work systems or practices taking place, four steps are essential, namely:

1 building acceptance
2 developing procedures
3 training management and operators
4 linking to other organizational systems.

The first stage entails obtaining the acceptance of senior management of the need for change. Without their approval, the change could be ineffective. At this stage, a team should be formed, its principal objective being to implement the transition from the old system to the new one.

The implementation team will be charged with reviewing the proposed plan and recommending administrative policies. This includes setting the timetable, goals and measures for the programme, determining staff and line requirements, and coordinating the integration of the new system with all other human resource programmes. A key benefit of creating such a team is that it ensures line input and support from the start. Generally, people support what they help create.

Training should be directed to providing those managers responsible for using the new system with the self-confidence to put it into operation, the skills to do it satisfactorily and the motivation to succeed.

Effective system change relies on open, honest and consistent communication with employees. The only major obstacle to the success of a system change is a lack of ongoing maintenance. Most systems fail not because of inherent flaws, but because no one ensures that they are maintained and linked to all other programmes and systems. Failure to maintain systems results in staff reverting to former systems and/or rejecting the new one.

A new system is fully integrated when it is no longer seen as a programme, but as a standard feature of the business operation after there is full explanation of planned changes, when comments and suggestions have been sought before final approval, and when the system interfaces rather than interferes with all other organizational efforts.

2

Individual change

At the individual level, most people are resistance to change. On the other hand, people are continually changing as a result of increasing age, experience, skills acquired, differing environments and changed objectives. Conservatism in attitudes and beliefs generally increases with age.

In most cases, managers and others, such as health and safety practitioners, are directly involved with bringing about individual change. They have the task of convincing individuals that the change is needed because the old system or arrangement is no longer appropriate or safe. Change may be accelerated, for instance, if there has been a serious accident that people can readily identify with, new machinery and equipment has been introduced or new legal requirements relating to some aspect of health and safety have come into effect.

Various techniques for bringing about change should be considered.

- *environmental change* this entails changing the social or physical environment of people to the extent that the previous behaviour does not bring the rewards it did formerly or, conversely, the new modified behaviour provides greater rewards.

- *information change* the provision of new or modified information to a person can show that the existing adaptation is not as close as the individual thought it was, or as it could be. A further approach is to improve communication skills so that people are more open to influence, such as sensitivity training.

- *goal or objective change* this involves changing the goals or objectives that an individual aspires to through the use of propaganda, training or various media or advertising campaigns.

CONCLUSION

All organizations are different. However, factors such as management style and culture are particularly significant in promoting and maintaining appropriate levels of health and safety performance.

The organization is a major influence in people's lives and determines their behaviour at work. As such, it is essential that the organization provides leadership in the field of health and safety, as opposed to the classic approach of 'sitting back and waiting for it to happen'. Leadership implies all the factors mentioned earlier – demonstration of commitment from the top, appropriate policies, the maintenance of an appropriate safety culture, clear identification of group and individual responsibility, and a system for ensuring that the policies approved by the Board are translated into effective action at local level.

The job

INTRODUCTION

A common outcome of the investigation of workplace accidents is to write the accident off as being solely caused by 'human carelessness', 'human error', 'human failure' or 'lack of attention' on the part of the accident victim. In many cases, this is the quick and easy way out for the investigator, implying that no further action is required to prevent recurrences.

Management of human factors

Successful management of human factors and the control of risk involves the development of systems designed to take account of human capabilities and fallibilities. On this basis, tasks should be designed in accordance with ergonomic principles, which allow for limitations in human performance. Matching individuals to jobs, both physically and mentally, will ensure that they are not over loaded and that they make the most effective contribution to the enterprise.

Physical matching includes the design of the whole workplace and working environment. *Mental matching* involves individuals' information and decision-making requirements, as well as their perceptions of risk. Mismatches between job requirements and workers' capabilities provide the potential for human error and, to some extent, the degree of compliance with health and safety practices.

Compliance with health and safety practices

There are a number of factors that influence compliance with health and safety practices, including:

- the interaction of individuals with their job and the working environment
- the influence of equipment and system design on human performance.

The design of the job and working environment should be based on *task analysis* of the actions required by the operator. From a health and safety viewpoint, task analysis provides the information for evaluating the suitability of machinery, tools and equipment, work procedures and patterns, and the operator's physical and social surroundings (see later in this chapter).

Equipment and system design is a basic feature of ergonomic study. Mismatches between the machine and the operator in the interpretation of displays and operation of controls is a frequent cause of accidents.

JOB DESIGN

Major considerations in job design include:

- identification and comprehensive analysis of the critical tasks expected of individuals and the appraisal of likely errors
- evaluation of required operator decision making and the optimum balance between human and automatic contributions to safety actions
- application of ergonomic principles to the design person–machine interfaces, including the display of plant and process information, control devices and panel layouts
- design and presentation of procedures and operating instructions
- organization and control of the working environment, including the extent of the workspace, access for maintenance work and the effects of noise, lighting and thermal conditions
- provision of correct tools and equipment
- scheduling of work patterns, including shift organization, the control of fatigue and stress, and arrangements for emergency operations/situations
- efficient communications, both immediate and over periods of time.

Safe systems of work

Section 2(2)(e) of HSWA places a duty on employers to provide and maintain plant and systems of work that are, so far as is reasonably practicable, safe and without risks to health. A safe system of work can be defined as 'the integration of people, machinery and materials in a correct environment in order to provide the safest possible working conditions'.

Safe systems of work, increasingly, are being laid out in written form and are used in training activities in addition to ensuring safe working. They can take several forms, such as permit to work systems and method statements, and are generally designed as a result of specific techniques, such as job safety analysis, which is derived from *task*, or *job, analysis*.

Task, or job, analysis

Task, or job, analysis is a characteristic procedure in industrial psychology and is when an occupation is studied to devise methods for the selection and/or training of employees. The analyst seeks to investigate the actual work performed and analyse it into various jobs or tasks. It should be appreciated that, while there are several variations of the technique, above all, the methods used should be comprehensive and systematic. The result of task analysis are commonly used for the determination of pay and in human resource planning.

Broadly, task analysis is concerned with the identification and, perhaps, assessment of the skill and knowledge components of jobs. It may also be used in the appraisal and description of person–machine relationships. Task analysis frequently incorporates *job training analysis*, which is defined as 'the identification and specification of what has to be taught and learned for individual jobs'.

Job safety analysis

Job safety analysis (or job hazard analysis) can be defined as the identification of all the accident prevention measures appropriate to a particular job or area of work activity and the behavioural factors that most significantly influence whether or not these measures are taken'. It can be seen, in most cases, as an extension of task or job analysis.

The job safety analysis approach is both diagnostic and descriptive. It may be job-based, for example, looking at the jobs of machinery operators or fork-lift truck drivers, or activity-based, such as looking at manual handling operation, roof work or cleaning activities.

Fundamentally, job safety analysis evolved from the techniques of:

- task analysis

- method study and work measurement.

It is principally used in the design of safe systems of work, in the preparation of job safety instructions and for job safety training purposes.

The **SREDIM** principle is commonly used, namely:

S = *select* the work to be studied
R = *record* how the work is done
E = *examine* the total situation
D = *develop* the best method for doing the work
I = *install* the method in the company's operations
M= *maintain* this defined and measured method.

Applying the SREDIM principle to job safety analysis, the procedure is as follows:

1 *select* the job to be analysed
2 break the job down into component parts in an orderly and chronological sequence of job steps and *record* these steps
3 critically observe and *examine* each component part of the job to determine the risk of accident
4 *develop* control measures to eliminate or reduce the risk of accident
5 formulate written safe systems of work and job safety instructions and *install* these within the organization
6 *maintain* the safe systems of work, reviewing them at regular intervals to ensure correct and continuing utilization.

Job safety analysis commonly takes place in two distinct stages.

1 *Initial job safety analysis* This stage examines a number of features of the job, namely:

- job title   department/section
- job operations   work equipment used

- materials/substances used protection required
- hazards degree of risk
- work organization specific tasks

2 *Total job safety analysis* At this stage, job safety analysis should take into account the following:

- *the operations* This should include the step-by-step operations in a process or work activity from start to finish

- *the hazards* Clear analysis of the hazards that could arise, such as risk of burns, hand injury or inhalation of dust, should take place

- *the skills required* Here consideration should be given to both the knowledge required on the part of the individual to conduct the operation safely and aspects of their behaviour that could create hazards, and attention should also be paid to various skills required, perceptual skills in particular

- *external influences on behaviour* These influences can be categorized in terms of:
  - the nature of the influence, such as noise, that could mask warning signals
  - the source of the influence, such as machinery, that could require special operating skills
  - the activities involved, such as the machine-loading procedure, that may require a specific sequence of operations.

- *the learning method* What the training procedure is going to be that will ensure that the operating methods are correctly learned and understood by the trainees, which may entail both practical and theoretical training techniques and some form of test or examination as to individual competence.

Job safety analysis record
Once the analysis has been completed, it should be recorded (see Figure 3.1).

Criteria for selection of jobs for analysis
Clearly it would be impossible, let alone cost-effective, to undertake job safety analysis for all forms of job. However, the following criteria

JOB SAFETY ANALYSIS RECORD		
Job title		Date of job analysis
Department		Time of job observation
Analyst/reviewer		
Description of job		
Accident experience		
Maximum potential loss		
Legal requirements		
Relevant information (codes of practice, etc.)		
Sequence of job steps	Risks identified	Precautions advised
Safe system of work		
Review date		
Job safety instructions		
Training needs		
Signed	Date	
Department	Function	

● **FIG 3.1 Example of a job safety analysis record**

should be considered in selecting jobs for analysis:

- past accident and loss experience
- the maximum potential loss that could arise
- the probability or likelihood of accident(s) recurring
- the severity of injury, damage or loss that could result from faulty job operation and/or accidents
- the number of people exposed to the risks and the relative levels of competence
- current legal requirements, for example, Manual Handling Operations Regulations 1992
- the relative newness of the job.

Job safety requirements and instructions

These are derived from the job safety analysis and are used as a means of communicating the safe system of work to the operator. They can be produced as:

- a general manual covering all jobs
- specific cards attached to equipment and/or displayed in the work area.

Job safety requirements and instructions should feature as part of the specific health and safety training required by staff under HSWA and the MHSW regulations.

Job descriptions

A job description is, fundamentally, a statement of the significant characteristics of a job. The majority of people have some form of written job description or job specification. Such a document identifies the job title, the person or post to whom the job holder is responsible, the job content in terms of the tasks involved, the principal responsibilities and accountabilities and, in some cases, criteria for assessment as to satisfactory performance. Very rarely, however, do job descriptions incorporate health and safety-related elements.

To ensure that people at all levels appreciate their responsibilities for health and safety, such responsibilities and accountabilities should feature as a standard part of the job description, together with the criteria for assessing performance on a regular basis.

Performance appraisal

Various forms of performance appraisal and assessment operate within working organizations. Generally, such performance is assessed annually, commencing with the Board Chairman appraising the Managing Director. This process is then cascaded down the organizational hierarchy to operator level. It is standard practice for appraisees to agree future objectives with their immediate manager and for subsequent performance in the achievement of these agreed objectives to be assessed at the next appraisal. In organizations operating performance-related pay systems, performance is directly related to the reward structure of the organization.

How often, though, do managers and their subordinates agree health and safety-related objectives as part of this system? Where such a system does operate, what is the basis for assessing performance? The use of departmental accident and sickness absence statistics as a measure of a manager's performance in this area, and the use of in-company accident 'league tables' cannot be said to be true measures of performance, largely due to the fact that individual managers do not have total control over these situations. Furthermore, such a negative approach can result in managers and supervisors actually discouraging the reporting of accidents, particularly if they are likely to be penalized later as a result.

A more positive approach to assessing health and safety performance on a unit or departmental basis would be by means of the review of reports of surveys, audits or inspections undertaken by the organization's health and safety practitioners. These reports would indicate progress achieved by individual managers in meeting legal and company standards, indicating a willingness or otherwise to improve health and safety performance.

HUMAN RELIABILITY

Increasingly, attention is being paid to human error, incorrect human action or the failure to act by people as causative factors in accidents of varying severity. In 1987, the Advisory Committee on the Safety of Nuclear Installations established a study group that was to report to the Committee on the part played by human factors in the incidence of risk in the nuclear industry and the reduction of this risk. The second report of the study group, 'Human Reliability Assessment – A Critical Overview', provides a comprehensive insight into the various classes of human error, the factors that contribute to human error and the technique of *Human Reliability Assessment*.

Human Reliability Assessment (HRA)

HRA includes the identification of all the points in a sequence of operations at which incorrect human action, or the failure to act ('sins of omission'), may lead to adverse consequences for plant and/or people. HRA techniques assign a degree of probability on a numerical scale to each event in a chain then, by aggregating these, arrive at an overall figure indicating the probability of human error for the whole chain of events. Such an assessment may point to the steps that need to be taken to reduce the likelihood of error at certain points by introducing organizational, procedural, ergonomic or other changes.

Classifications of kinds of human error

There are several kinds of human error.

- *Unintentional error* This may arise where an individual fails to perform a task correctly, such as operating a control or reading a gauge. These are typical slips or lapses frequently associated with carelessness or lack of attention to the task.

- *Mistakes* In this situation, the individual shows awareness of a problem, but forms a faulty plan for solving it. They will thus carry out, intentionally but erroneously, action(s) that are wrong and may entail hazardous consequences. Typical examples are in the operation and maintenance of plant and machinery, and in assembly work. Failure to correct individual mistakes through training and supervision can lead to disaster situations.

- *Violations* In this case, a person deliberately carries out an action that is contrary to some rule which is organizationally required, such as an approved operating procedure. Deliberate sabotage is an extreme example of a violation. Violations involve some complicated issues concerning conformity, communications, morale and discipline. In piece-work systems, the removal of machinery guards or overriding safety mechanisms in order to increase output and, therefore, remuneration is a common violation.

- *Skill-based errors* These arise during the execution of a well-learned, fairly routine task. They are amenable to prediction, either from laboratory experiment or from experience in other skilled tasks, even when the tasks are performed in different industries. For instance, the probability of a skilled typist striking the wrong key depends on the nature and complexity of the material being keyed, but, on average, the error rate turns out to be much the same in an office of any kind, as in laboratory experiments.

Skill-based errors commonly occur among the more highly skilled members of the organization. Such people, because of their acquired skills and experience in the work process, frequently take the view that the safety precautions imposed are purely for the benefit of the underskilled and underexperienced operator. This is a classic case of familiarity breeding contempt. Such an attitude to work is accompanied by overconfidence in the task, which can lead to, particularly, machinery-related accidents.

- *Rule-based errors* These are errors that occur when a set of operating instructions or similar set of rules is used to guide the sequence of actions. They are more dependent on time, on individual cast of mind and on the temporary physical, mental or emotional state of the person concerned. They are, therefore, harder to predict from past statistics or from experiment.

- *Knowledge-based errors* Also called *errors of general intention*, these are errors that arise when a choice decision has to be made between alternative plans of action. They arise when the detailed knowledge of the system possessed by the person and the resulting mental model of it is incorrect. Such errors, which are extremely difficult to predict, can only be forecast by an analyst who possesses the insight to predict the form of this model.

CONCLUSION

Human capability and the potential for error are important factors in the selection of people for different tasks. Tasks should be designed using techniques like job safety analysis, with particular attention being paid to the various influences on behaviour that the task produces.

Techniques based on Human Reliability Assessment will predict the potential for human error, and, linked with ergonomic principles in the design of work layouts, systems and machinery and vehicle controls and displays, should be used in the process of job design.

3

4

Personal factors

Personal factors, such as attitude, motivation, perception, personality, training and the potential for human error, are significant in any consideration of human factors and safety.

Clearly, all individuals are different from each other, particularly in terms of their effectiveness in undertaking a task. These individual differences may be associated with characteristics passed on from their parents (inherited characteristics) and from a broad range of life experience situations, all of which are specific to each individual.

The differences between people

A number of factors contribute to the differences between people, particularly in terms of individual behaviour. These can be classified thus:

- genes and chromosomes
- experiences in the womb
- birth trauma
- family influences:

- cultural patterns of child rearing
- the effect of one's parents in terms of affection received, their personalities, time devoted to the individual, the types of interaction that have taken place in the past, the level of strictness and so on
- the influence of brothers and sisters

- socio-economic group
- position in the family
- general family background

- geographical location
- pre-school influences – people, situations
- education received – quality, support given, opportunities available
- occupational factors:

- training and retraining received
- the status of the actual occupation and position held
- the opportunities available in the particular occupation
- the effects of membership of a particular work group

- hobbies and interests
- specific family influences, such as the effects of marriage, children and family life
- the effects of the ageing process.

All these factors, and probably many more, have the effect of producing a unique individual who has their own particular views about a range of issues, including health and safety at work.

On this basis, the selection process for a particular job should incorporate both the specific aspects of the job (job description) and the characteristics of the individual that make them best fitted to undertake the job (personal specification).

The seven-point plan

Alec Rodger classified these specific characteristics into seven categories, thus:

1 *physique (health and appearance)*:
- height
- build
- hearing
- eyesight
- general health

- looks
- grooming
- dress
- voice

2 *attainments*::
- general education
- job training
- job experience

3 *general intelligence*:
- tests
- general reasoning ability

4 *special aptitudes:*
- mechanical
- manual dexterity
- skill with words
- skill with figures
- artistic ability
- musical ability

5 *Interests*:
- intellectual
- practical constructional
- physically active
- social
- aesthetic

6 *Disposition*
- acceptability
- leadership
- stability
- self-reliance

7 *Circumstances*
- age
- marital status
- dependants
- mobility
- domicile.

Other points that may be significant in the working situation can be added to this list depending on the type of work to be undertaken. Problems can arise, however, in that certain factors are easier to specify and evaluate than others, and other factors, such as personality, perception, attitude, motivation and the ability to process information, need to be considered at the same time.

ATTITUDES

Attitudes are an important feature of human behaviour. Many definitions of the term 'attitude' have been put forward over the years, such as:

'a predetermined set of responses'

'a specific mental disposition toward an incoming (or arising) experience, whereby that experience is modified, or a condition of readiness for a certain type of activity'

'the disposition of people to view things in certain ways and to act accordingly'

'a mental disposition of the human individual to act for or against a definite object'

'a shorthand way of responding to a particular situation'

'a tendency to respond in a given way in a particular situation'

'a more or less stable predisposition or readiness to react in a positive, negative or neutral way to different things, people or situations'

'a mental and neural state of readiness, organized through experience, exerting a directive or dynamic influence upon an individual's response to all objects and situations with which it is related'

'a learned orientation, or disposition, toward an object or situation which provides a tendency to respond favourably or unfavourably to the object or situation'. **4**

Attitudes comprise a *cognitive* component and an *affective* component. The cognitive component is concerned with thoughts and knowing, such as perceiving, remembering, imagining, conceiving, judging, reasoning, the analysis of problems and decision-making processes. The affective component, on the other hand, is concerned with emotions or feelings of attraction or revulsion.

The functions of attitude

According to D. Katz, there are four functions of attitude.

1 *Social adjustive function* This is concerned with how people relate, and adjust, to the influence of parents, teachers, friends, colleagues and their superiors. It is argued that by the age of nine years, most attitudes are established. Behaviour is based, to some extent, on a philosophy of 'maximum reward', minimum punishment'.
2 *Value expressive function* Individuals use their attitudes to present a picture of themselves that is pleasing and satisfying to them. This is an important factor, in that people see themselves as better and different, in some special way, from others around them (self-image). To promote this self-image, people may adopt extreme views of sit-

uations, dress in a particular way and adopt a particular political persuasion.

3 *Knowledge function* Attitudes are used to provide a system of standards that organize and stabilize a world of changing experiences. On this basis, people need to work within an acceptable framework, have a scale of values and generally know where they stand.

4 *Self-defensive function* This is concerned with the need to defend one's self-image, both externally, in terms of how people react towards us, and internally, to deal with inner impulses and our personal knowledge of what we are really like.

Attitude formation and development

Attitudes are formed as a result of continuing experience of situations during a lifetime and are difficult to change. They are directly associated with:

- self-image – the image that an individual wishes to project to the outside world, for example, affluent, stern, well-mannered, fair-minded

- the influence of groups, and group norms, that is the standards upheld by a particular group, say, a professional body, and whereby membership of the group entails sharing their attitudes and conforming with the norms

- individual opinions, including superstitions, such as 'All accidents are Acts of God', implying that nothing can be done in terms of preventing accidents.

Attitude change

Attitudes are difficult to change. In many cases, people simply do not wish to change their attitude to a particular situation, despite overwhelming evidence to support such a change. To be successful, attitude change must take place in a series of well-controlled stages. First, by attracting the attention of the individual to the fact that a change of attitude is needed and, second, by convincing this person that their current attitude is inappropriate or wrong.

Cognitive dissonance

One of the barriers to attitude change is *cognitive dissonance*, the situation of conflict that results, when a person holds an attitude that is incompatible with the information presented. The theory of cognitive dissonance was proposed by Leon Festinger, who postulated that, when faced with two pieces of information – *cognitions* (knowledge, thoughts, feelings) – that are inconsistent, opposite or conflicting, people will feel uncomfortable. They will, therefore, wish to reduce this discomfort (dissonance) by engaging in a variety of activities, such as changing their ideas, beliefs, knowledge or skills, or by avoiding the thoughts altogether.

People frequently display cognitive dissonance when required to work in a particular safe way that may be new to them, to use a specific safety device or when required to use some form of personal protective equipment. Where cognitive dissonance may be encountered among a group of workers, the ideal attitude change process must take place as a series of specific considerations by each individual in the group. Thus, in the situation where there may be risk of eye injury and there is a need for operators to wear eye protection, the ideal sequence of thought on the part of operators should be:

4

1 *'I do not wear eye protection. I'm a skilled operator!'*
2 *'People have been blinded using that equipment in the past.'*
3 *'I suppose I could be blinded when using that equipment'.*
4 *'If I wear the goggles all the time, I should avoid being blinded.'*
5 *'I'll wear the goggles in future!'*

The above is the ideal attitude change sequence. However, it is affected by factors such as past experience, understanding of the risks involved and the seriousness of the potential injuries.

Factors in attitude change

There are many factors that must be considered when endeavouring to bring about changes in attitudes, particularly with regard to safe working practices.

1 The individual

• *Built-in opinion* An opinion is a statement of something that can be subject to change. People frequently acquire a particular opinion or viewpoint during on-the-job training, particularly where the trainer is unaware of the inexperience of the trainee in terms of safe working. Conversely, opinions can harden with the ageing process becoming progressively more difficult to change as people get older. In certain cases, people may be prepared to change their opinion where some direct benefit is identifiable.

- *Conservatism* Most people are conservative in their views and become increasingly resistant to change. The degree of conservatism is a specific feature of the individual and their attitude to safe working, and may be a significant barrier to attitude change.

- *Past experience* Most people carry out tasks on the basis of past experience, using working practices they have picked up over the years. The majority of people learn by their mistakes and modify their behaviour in order to prevent repetition of mistakes.

- *Level of intelligence and education* These factors are important elements in framing attitudes. Well-educated and intelligent people may consider, for instance, that the safety precautions laid down are directed principally at those with poor intelligence and education and do not apply to themselves. Furthermore, they may feel that their intelligence will automatically protect them from hazards in the workplace.

- *Motivation* Motivation, the element of human behaviour that drives people forward, can have a direct effect on attitudes to safe working. Bonus and incentive schemes are commonly directed at improving the motivation of workers to greater levels of productivity, thereby reaping the rewards offered if targets are achieved. On this basis, operators may be motivated to take risks if the benefits of taking these risks are readily identifiable.

- *'The sheep index'* This is the extent to which some people will blindly follow those whose views and opinions they respect and value without question. As such, they take on board the attitudes of their 'leaders', giving little or no thought as to the correctness or otherwise of these acquired attitudes. This factor can be significant where there may be opposition from workers to, for instance, the introduction of a safe system of work or an amendment to an existing system.

- *Credibility* Attitudes are directly associated with the beliefs that an individual may have. Thus, we are more likely to consider attitude changes suggested if the person making such suggestions has some form of credibility. Credibility can be associated with an individual's rank or level of prestige within the organization, so that people are more likely to consider changes in attitude to, for instance, safe working practices if the messages are seen to be coming from a senior manager or director than, perhaps, a supervisor.

- *Attractiveness* Any change in attitude requested must be attractive to the individual concerned. Attractiveness incorporates three elements: similarity, friendship and liking. The more similar two people see themselves to be in terms of, for instance, a particular work situation, the more likely they are to believe each other. Friendship is also an important factor in attitude change, in that a person is more likely to take notice of a friend than someone seen as hostile. Similarly, where an individual actually likes and gets on well with their supervisor, they are more likely to change their attitude to safe working to maintain this situation.

- *Selective interpretation* There is an old saying, 'People only hear what they want to hear'. Whether or not a message gets through to bring about an attitude change in the desired direction depends, in large measure, on how the receiver interprets the message. In many cases, they are very likely to interpret the facts presented by selecting those that fit in with their existing attitudes. This is especially true where there is a large discrepancy between their attitude and the message. Selective interpretation of facts is one of the potent factors in freezing attitudes so that they resist change.

- *Immunization* This aspect can be compared with the medical practice of immunizing an individual against a particular disease by inoculating them with a controlled dose of the organism that causes the disease. In the case of attitudes, the analogy means that a mild exposure to an opposing attitude can immunize a person against it so that they will never accept further facts or arguments for it, no matter how strong they are.

2 Attitudes currently held

- *Cognitive dissonance* The problem of cognitive dissonance is a well-established barrier to attitude change and must be considered in the design of training programmes and other activities directed at changing attitudes.

- *Self-image* Attitudes currently held are an important feature of an individual's self-image. Similarly, self-image may be identified with a specific set of attitudes. In many cases, an individual may fear loss of face or self-image if they change their attitude to a particular matter. As such, self-image can represent a further barrier to attitude change, and must be taken into account in measures to improve or change attitudes.

- *Group norms* Groups of workers develop their own specific norms. These are the rules or standards that specify appropriate and inappropriate behaviour of people in standardized situations. Thus, membership of a particular group entails conformance to such standards. For training exercises to be effective, they should involve the complete working group and examine current group norms with a view to obtaining a consensus on any necessary changes in attitude.

- *Financial gain* There is no doubt that the prospect of financial gain can have a significant effect on attitudes held, if only for a short period. People are known to modify their behaviour generally if some form of reward is offered. However, this short-term change in attitude can be lost when the reward is removed, people frequently reverting to their original attitudes.

- *The opinions of others* As stated earlier, an opinion is simply defined as 'a statement of something which is highly subject to change'. An opinion lies somewhere between an attitude and a belief. People generally value the opinions of others for whom they have respect, say, a relative, manager, shop steward, safety representative.

- *Skills available* The skilled operator will frequently hold the attitude that their skills, acquired over many years, will automatically protect them from accidents or adverse situations. Such an attitude can be dangerous in that it leads to complacency in observing safety procedures. In such a case, it may be necessary for operators with similar skills and experience to draw this person's attention to the fact that a change of attitude is required.

3 The situation

- *Group situations* Where people work together as a group, they can develop a group attitude towards safe working practices that may be inappropriate. Any training directed at changing attitudes should therefore take place on a group basis.

- *The influence of change agents* Change agents include managers, enforcement officers, health and safety practitioners and officers of insurance companies. Enforcement officers can take action whereby various sanctions can be imposed by the courts. In certain cases, such sanctions may bring about changes in attitude, on the part of employers particularly, but this change in attitude may be limited.

  The role of health and safety practitioners varies substantially from one organization to another. They generally have only an advisory function and the person receiving the advice can accept it as received and take action on it, modify the advice or reject it. In order to bring about changes in attitude at local level, the health and safety practitioner's role must be clearly defined and supported by senior management.

  Insurance companies, on the other hand, tend to take a more negative post-incident role that is based, primarily, on the investigation of claims made by injured employees. Increasing an organization's insurance costs, such as employer's liability, may bring about some degree of attitude change on the part of management towards the implementation of safe working practices.
- *Prestige* Increasingly, organizations are seeing health and safety and the procedures they adopt as a means of gaining prestige in the market-place. The philosophy that 'a safe company is a profitable company' is increasingly finding favour as a means of promoting the image of the organization. Whether or not this viewpoint necessarily brings about attitude change further down the company hierarchy is a matter for conjecture, but the integration of health and safety with quality management, for instance, has gone some way to improving attitudes to the subject in many organizations.
- *The climate for change* Most organizations go through periods of change. The climate for change towards better and safer working practices, together with the establishment and development of a safety culture, linked with increased personal accountability for health and safety on the part of managers, can bring about changes in attitude at all levels provided, of course, the effort is sustained.

4 **Management example** This is the strongest of all motivating factors in getting people to improve their attitudes to health and safety at work. Management example, for instance in the wearing of personal protective equipment or in following a particular safe system of

work, must be maintained on a continuing basis. Poor management example, on the other hand, can result in loss of credibility of the safety procedures and systems.

5 **Conflict of characteristics** A person who is endeavouring to persuade another to change their attitude may have some characteristics that are favourable to attitude change and some that are not favourable. For instance, an older person may have credibility because of their qualifications and length of experience. On the other hand, they may not be so attractive to younger people as other young people. This can sometimes result in a conflict of characteristics in the source of the message.

6 **Appeals to fear** A common technique in persuasive messages aimed at changing attitudes is to make the receiver fear that something bad or undesirable could result if they disregard the message. Classic examples of this technique are seen in the 'Don't drink and drive campaign' posters and television campaigns over the last 20 years. As with the use of horror posters, it is debatable whether any longstanding change in attitude takes place.

MOTIVATION

The term motive implies a need and the direction of behaviour towards a goal, aim or objective. A motivator, on the other hand, is something that provides the drive to produce certain behaviour or to mould behaviour. For instance, for many years, physical punishment was seen by the authorities as a means of moulding the behaviour of a whole range of people – school children, members of the armed forces, prisoners.

The word motivation is used to describe the goals or objectives that people endeavour to achieve and the drive or motivating force that keeps them on track in pursuing these goals or objectives.

The word motivation, then, can be used to refer to states within an individual, to behaviour and to the goals towards which behaviour is directed. Motivation has three specific features:

• a motivating state within the individual

- behaviour aroused and directed by this state
- a goal or objective towards which the behaviour is directed.

When the objective is achieved, the state that caused the behaviour subsides, thus ending a cycle until the state is aroused again in some way. This cyclical process is shown in Figure 4.1.

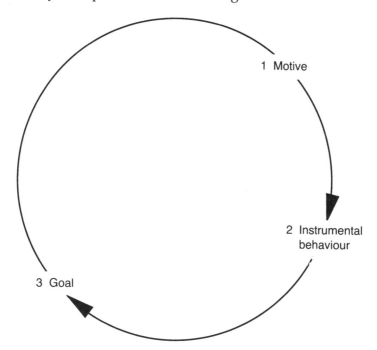

1 Motive

2 Instrumental behaviour

3 Goal

- **FIG 4.1 The motivational cycle**

Theories of motivation

Many theories of motivation have been directed at identifying why people work. Several of these theories are outlined below.

F. W. Taylor (1911): theory of the management and organization of work

F. W. Taylor examined the differences between potential managers, who were involved in the organizing, planning and supervising of work operations, and the rest of the workforce, who were not particularly interested in these aspects. They preferred to have simple tasks organized for them that they could be trained to carry out without having to make decisions. Once the work had been planned, money was the principal incentive for increasing production.

Taylor said, 'Man is a creature who does everything to maximize self-interest'. In other words, people are primarily motivated by economic gain. 'What's in it for me?' is seen as an important motivator for some people.

This viewpoint, that people only went to work for money, was held by many of the entrepreneurs of the nineteenth and early twentieth centuries. The philosophy that the more money a person was paid, the harder they would work, led to a whole range of variations on this theme – incentive schemes, productivity bonuses, piece-work, payment by results and even the concept of paying 'danger money' for high-risk work activities. Taylor's concept of *Scientific Management* brought about basic strategies and systems, such as the division of labour, mass production processes, work study and great emphasis on optimum conditions for work, training and selection of staff.

This theory has now largely been discredited. While everyone recognizes that financial rewards are important in order for people to survive and maintain a reasonable standard of living, other factors are more important.

Elton Mayo (1930s): 'Social Man'

Various studies by Elton Mayo at the Hawthorne Works of the Western Electric Company, Chicago, in the 1930s endeavoured to identify the best working conditions for people with a view to improving productivity. Former attempts to improve worker performance and thereby increase output were based on notions that workers should be regarded as machines, whose output could be improved by attention to the physical conditions and the elimination of wasteful movements and the resulting fatigue.

Mayo's methods involved studying the output of a small group of female workers who were consulted about each of a large number of changes in working practices introduced. Nearly all these changes led to improvement in output. However, when these continuing changes eventually returned to the original physical conditions, output increased still further. Such revelations led Mayo to conclude that, as the changes observed could clearly not be attributed to the changes and physical conditions as such, they must be due to social factors and to a change in the attitude of the workers. Because someone was clearly showing an interest in the problems of these workers, they experienced a feeling of importance and responded accordingly by giving their best. It was found, for instance, that most grievances raised were only symptoms of a general discontent, and that a chance to discuss these grievances in the presence of a sympathetic listener frequently led to solving them.

The need for their cooperation, and the fact that consultation had taken place over changes, resulted in the formation of a group culture. Members valued being in the group, felt responsible to the group and the company and, particularly, were concerned with meeting the standards of performance established by the group. The outcome of these studies was the realization that people were not directly motivated by the financial rewards, but tended more towards a social philosophy of 'a fair day's work for a fair day's pay'.

They were particularly concerned with the social interaction that took place during work periods and took pride in belonging to an identifiable work group. These factors alone provided a high level of job satisfaction, the more significant the work group, the greater the satisfaction. They also responded to pressure from their peers and the interest shown in them individually by various levels of management. On this basis, work was seen as a social activity, not just a means of obtaining pay.

A number of important points emerged from Mayo's studies into worker attitudes between 1927 and 1932, in particular:

- the opportunity to air grievances had a beneficial effect on morale

- complaints were not often objective, but symptomatic of a more deep-rooted disturbance

- workers are influenced by experiences outside their workplace, as well as those inside the workplace

- dissatisfaction was often based on what the worker saw as an underestimation of their social status in the firm

- voluntary social groups formed at the workplace had a significant effect on the behaviour of individual members, giving rise to group norms for production and group sanctions for people who went outside these norms or established informal standards.

Mayo's studies further resulted in a changed emphasis on the role of the supervisor from that of autocrat to group leader, and a greater attention to the need for establishing and maintaining the morale of work groups.

Abraham Maslow (1943): 'Self-Actualizing Man'

Maslow studied the factors significant in the motivation of successful people and that gave them satisfaction in their work. Achievement, self-esteem and personal growth featured strongly in the outcome of these studies.

Maslow defined motivation as 'a striving to satisfy a need'. Thus, when an individual needs something, they will strive to satisfy this need. As a need is satisfied, it uncovers another level of needs, so people go on wanting. Maslow actually categorized these needs and ranked them in order of importance, producing a hierarchy of needs (see Figure 4.2) thus:

- *basic or physiological needs* such needs as those for food, drink, fresh air, warmth, sleep, rest, shelter, sex
- *safety and security needs* the needs for freedom from physical and mental attack, deprivation, insecurity and want; the need for physical and psychological safety and security, for shelter
- *social needs* those needs associated with status and acceptance as a member of the group, that is, a sense of belonging, participation in social activities, self-esteem, respect from one's peers, the need to love and be loved
- *self-esteem (ego) needs* the need for achievement, praise, honour, glory and acclaim, the need to be noticed, the need for self-respect, status and the respect of others, for competence, knowledge, independence, responsibility, self-confidence and prestige
- *'self-actualization'* the ultimate stage of achieving one's potential, 'doing your own thing', 'what a man can be, he must be!' – self-actualization is synonymous with growth, personal development, accomplishment, self-fulfilment, self-expression and creativity.

Maslow stated that, at any point in time, people are progressing from their basic needs to a higher level of need or acting according to a current level of need. When, and only when, a lower need has been satisfied, the next highest becomes dominant, and the individual's attention is directed to satisfying this higher need. At the higher level, the need for self-actualization is destined to remain unsatisfied as new meanings and challenges arrive. All the above categories of need remain permanently active. What varies is the strength or degree of influence of each of these needs at any one time.

● **FIG 4.2 Maslow's hierarchy of needs**

The principal outcome of Maslow's studies is the fact that *all* needs must be studied in an attempt to motivate people at work. Attempts to provide motivation through the use of incentives relevant to the needs of a lower or higher level, and not that need which is currently dominant, are likely to fail. Furthermore, individuals and groups differ with regard to needs that are dominant. On the basis of this, the same incentives cannot be expected to motivate everyone, so management should endeavour to establish the dominant needs and act accordingly.

Fred Herzberg (1957): Two Factor Theory

Fred Herzberg undertook an extensive study throughout the United States of America, Canada and the United Kingdom, seeking to identify among workers in many organizations the factors that produced job satisfaction and dissatisfaction.

Herzberg examined homeostatic needs (*hygiene factors* or *maintenance factors*), which are concerned with avoiding pain and dissatisfaction, and growth needs (*motivators*), which are concerned with actively seeking and achieving satisfaction and fulfilment.

He asked many people in different jobs at different organizational levels two questions:

- What factors lead you to experience extreme dissatisfaction with your job?
- What factors lead you to experience extreme satisfaction with your job?

He established that there was no one factor that determined the presence or absence of job satisfaction. However, he did point out that before satisfaction with work can be improved, the factors that cause dissatisfaction must be dealt with.

In the typical work situation, maintenance factors involve the total environment affecting the employee–physical conditions, pay, safety, security, social factors and interpersonal relationships. On the other hand, motivational needs that lead to positive happiness are the needs for growth, achievement, responsibility, accountability and recognition. These needs can only be met by undertaking the actual work itself. Herzberg's view was that the job itself can provide a potentially more powerful motivator than any externally-introduced incentives. Employees can be actively satisfied only when the work done is perceived by the worker as being meaningful and challenging, thereby fulfilling their motivational needs.

The hygiene factors and motivators are listed below:

hygiene factors:	*motivators*
– money	– challenge
– working conditions	– responsibility
– safety arrangements	– advancement
– quality of supervision	– interest and stimulation created by the job
– administrative procedures	– achievement
– interpersonal relationships	– recognition
– status	– possibility of growth.
– security	

People expect the hygiene factors to be present and satisfactory. If absent or poorly managed, they will bring about dissatisfaction. On the other hand, while motivators give rise to job satisfaction, they will not result in dissatisfaction if absent.

The message that comes from Herzberg's studies is that if management is to motivate people to take greater responsibility and stimulate job interest, it must get the hygiene factors, including health and safety arrangements, correct first.

4

Improving motivation

Herzberg suggested the concept of *job enrichment* as a solution to the problem of meeting motivational needs. Care must be taken, however, to ensure that job enrichment techniques are not introduced in such large amounts or at such speed as to arouse excessive alarm or fear among workers. When introduced on a planned basis, and with consultation, job enrichment makes it possible for the growth and achievement needs of workers to be met as a result of their efforts at work.

The significance of the job enrichment concept lies in the clarity with which it focuses attention on the motivational distinction between:

- task and environment
- intrinsic and extrinsic factors (see Figure 4.3)

According to Herzberg *task impoverishment* – namely removing the individual interest, challenge and responsibility from a job – results in

THE TASK

Motivators –
leading to
individual
and group
satisfaction.

Opportunities for
achievement.
Recognition of achievement.
Interesting and challenging
work.
Scope for the individual.
Advancement and growth.

**Organizational
efficiency
and
effectiveness**

THE ENVIRONMENT

Maintenance factors –
hygiene factors
that can help
overcome
dissatisfaction.

The organization policy.
Technical supervision.
Working conditions.
Pay.
Human relationships.

● **FIG 4.3 Job enrichment**

deteriorating motivation. Conversely, no amount of environmental improvement can compensate for this task impoverishment. Clearly there is a need therefore, to examine the tasks that people do with a view to identifying the factors that provide interest, challenge and responsibility for workers. These higher levels of motivation can be achieved through job enrichment, job enlargement and job rotation, accompanied by various worker participation schemes:

● *job enrichment* an increase in satisfaction and the responsibility attached to a job is achieved either by reducing the degree of super-vision or by allocating each worker a unit of work in which they have freedom to select their method and sequence of operations

● *job enlargement* in this case, the worker is required to progressively increase the actual number of operations that they undertake

● *job rotation* while the concept of job rotation has, in some instances, been unpopular with workers, the objective is to give more variety on simple, repetitive and usually automated tasks

● *worker participation* worker participation in varying degrees can be achieved in the following areas:

– personnel-related decisions – transfers to other jobs, disciplinary matters, various forms of training and instruction

– social decisions – welfare arrangements, health and safety procedures and systems of work, regulation of working hours, rest periods
– economic decisions – methods of production, production planning and control, production times, rationalization, changes in plant organization, expansion or contraction.

The fundamental objective is to reduce or eliminate the authoritarian approach to management and replace it with a more participative style of management. The participative leader is one who plans work and consults their subordinates as to the best course of action. They are skilled in reconciling conflicts so as to achieve group cohesion and effectiveness. Such a person is also interested in individual employees and their particular problems, whereas the authoritarian uses rewards and punishment of the traditional sort, exercises close supervision and is more interested in the activities of those above than those below them.

4

Motivation and safety

Important factors for consideration in motivating people to improved levels of safety performance include the following.

● *Joint consultation* Consultation with operators in planning the work organization is perhaps one of the greatest motivators from a health and safety viewpoint. Consultation is best undertaken through a health and safety committee with clearly defined objectives that is representative of all parties concerned. Such a committee should meet regularly, publish agendas and minutes, implement its decisions consistently and in an expeditious manner. Above all, it must have credibility with the workforce. Trade union safety representatives have, further, an important role to play in the joint consultation process. It is important that the role and functions of such persons are clearly identified, and that they are adequately trained in the various aspects of occupational health and safety in order to make as constructive a contribution as possible.

   In certain cases, small working parties can examine a particular situation, reporting their findings to the committee.

● *Attitudes held* The attitudes held by both management and workers are true indicators of the importance attached to health and safety. It is important that both groups be positively motivated towards improving standards of performance.

- *Communication systems* Communication systems within the organization should provide information that is comprehensible to all concerned. Many people become demotivated by communications that they find difficult to understand or interpret.

- *Quality of leadership* Leadership, as with all areas of management, should come from Board level if people are to be adequately motivated towards improved standards of health and safety performance.

Planned motivation schemes

Planned motivation is a technique by which the attitudes, and thereby the performance, of people at work can be improved. Such schemes are, in effect, an industrial catalyst – a tool to maximize performance – and are used in many organizations to improve the performance of, for instance, salespeople, engineers, line managers and others.

In the field of occupational health and safety, they mainly take the form of safety incentive schemes.

Safety incentive schemes

These are a form of planned motivation, the main objectives being that of providing motivation for people by:

- identifying targets that can be rewarded if achieved
- making the rewards meaningful and desirable to the people concerned.

Any planned motivation scheme should always be viewed carefully, however, due to the fact that such schemes may alter behaviour, but not necessarily attitudes.

From a health and safety viewpoint, safety incentive schemes have been found to be most effective when:

- people are restricted to one area of activity, such as work in a laboratory or workshop

- measurement of safety performance is relatively simple
- there is regular stimulation or rejuvenation
- there is support from both management and trade unions
- the scheme is assisted and promoted by means of the use of safety propaganda and training activities.

Individual needs and safety incentives

These aspects may be summarized as follows.

Motives:	May be satisfied by:
• financial gain through increased departmental or company profits	• monetary awards through suggestion schemes, profit-sharing plans, promotions, propaganda, increased responsibility
• fear of painful injury, death, loss of income, family hardship, group disapproval or ridicule, criticism by managers/supervisors	• visual material – posters, films, videos, public reports of accidents
• participation, that is, the desire to be 'one of the gang'	• group and individual activities – safety committees, working parties, safety campaigns
• competition – a desire to beat others	• health and safety competitions and award systems
• pride in safe workmanship, both individual and group	• recognition for individual and group achievement – trophies, awards, publicity
• recognition – desire for approval of others in group and family, for praise from supervisor.	• publicity – photographs and articles in company and community newspapers, use of notice boards.

Successful safety incentive schemes

In the design and operation of safety incentive schemes, the following points should be considered:

- they are best linked with some form of safety monitoring, whereby improvements can be measured and compared with previous performance, for example, safety sampling exercise, safety inspections
- correct, meaningful and achievable targets should be set
- on no account should safety incentive schemes be linked with acci-

dent rates at this can discourage the reporting of accidents

- if based on the lost time concept, how quickly people return to work after, particularly, an accident is significant – no two people with comparable injuries necessarily take the same amount of sick leave

- they tend to be short-lived and can get out of hand if not properly organized and monitored

- they must be accompanied by appropriate publicity and include feedback to staff on the progress in the scheme

- they can shift responsibility for health and safety from management to staff and this needs to be given careful consideration in the planning of a safety incentive scheme.

The 'quality of working life' philosophy

In 1974, the International Labour Office Conference passed a resolution directed at promoting and ensuring:

> protection against physical conditions and dangers at the workplace and its imme- diate environment; adaptation of installations and work process to the physical and mental aptitudes of the worker through the application of ergonomic princi- ples; prevention of mental stress due to the pace and monotony of work, and the promotion of the quality of **working life** [my italics] through amelioration of the conditions of work, including job design and job content and related questions of work organization; the full participation of employers and workers and their orga- nizations in the elaboration, planning and implementation of policies for the improvement of the working environment.

The *quality of working life* philosophy is one that takes the various theories of motivation and worker satisfaction and places them in a wider framework. It considers the numerous factors that affect the basic satisfaction and, hence, the motivation, of workers, and com- prise the quality of their life at work. It emphasizes the importance of consultation and worker participation as integral features for improv- ing the quality of working life.

It is generally felt that a well-satisfied and motivated worker is less likely to contribute to accidents to themselves or fellow workers. Fur- thermore, the potential for human error, which could result in acci- dents or the manufacture of defective products, is greatly reduced, and productivity is likely to be better. The provision and maintenance of a working environment, free from environmental stressors, such as excessive noise, poor levels of illumination and inadequate welfare amenity provisions, is an important feature in the quality of working life.

PERCEPTION

Perception is a general term referring to the awareness of objects, qualities or events stimulating a person's sense organs. Perception is an important feature of human behaviour in that people behave in accordance with the way they perceive situations, places, people and the world in general. It is a process of receiving information or inputs through the sensory channels – sight, hearing, touch, taste and smell – with the perceptual processes acting on this information to form it into the various areas of experience.

Perceptual experience is characterized by a number of features: **4**

- selectivity of input
- the organization of the particular input
- constancy of experience in spite of varying inputs
- a dimension of depth
- movement
- the effect of context
- the importance of the relationship between inputs
- the influence of both learning and motivation

How people perceive risk is associated with a number of behavioural factors – attitude, personality, memory, their ability to process information, the level of training received, the level of arousal and individual skills available. Perception of risk may further be affected by an individual's past experiences, the context in which the stimulus or information is presented and their current level of knowledge.

Thus, people perceive and gain information about the world around them by the use of their senses of vision, hearing, touch, taste and smell. Sight is the most important form of perception and is a significant factor in accident prevention and in the specific capabilities of individuals. Perception is, fundamentally, an active process of *attention* and *interpretation*.

The process of perception has been defined in a number of ways.

Perception is not determined simply by stimuli patterns. Rather it is a dynamic searching for a 'best' logical interpretation of the available data. Perception involves going beyond the immediately given evidence of the senses.

Richard Gregory, 1986

We do not perceive objective reality, but rather our construction of reality. Our sense organs gather information which the mind modifies and sorts. This 'heavily filtered' input is compared with memories, expectations, needs, attitudes, social conditioning, group control, until finally our conscious perception is 'constructed' and we have a 'best guess' at reality.

Ornstein, 1975

Perception is a process of assembling sensations into a useable mental representation of the world.

Coon, 1983

Perceptual set

The processes of attention and interpretation are closely related, and the factors that influence these processes, and their modes of operation, are frequently referred to as an individual's *perceptual set*. Perceptual set consists of:

● *external determinants* (the nature of the stimulating conditions):
– location
– senses and sense combinations
– intensity
– size
– colour
– tone
– motion
– novelty and change
– repetition
all of which can, and do, act in combination

● *internal determinants:*
– conscious or unconscious motives
– expectations
– capacity
– sensitivity
– span of attention

– change in attention
– fatigue
– needs (deprivation)
– culture
– sensitization and habituation
– prejudice.

These make up the perceptual set of an individual. To a greater or lesser extent, they regulate the impacts of the external determinants and how they are interpreted.

The above indicates that people have a perceptual set that selects, rejects, modifies, ignores or interprets that which is perceived. The various components of a perceptual set can be summarized as shown in Figure 4.4.

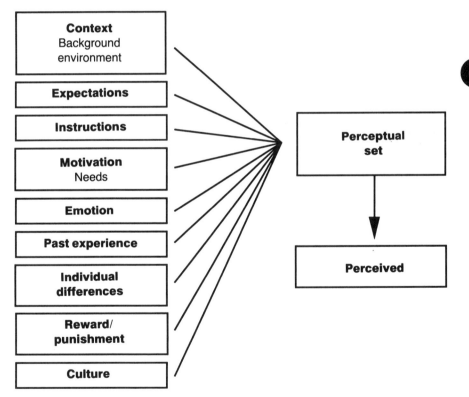

● **FIG 4.4 The components of a person's perceptual set**

Prejudice

Prejudice is when a specific stimulus is pre-judged, and this is based on a process of categorization of certain stimuli. It is a favourable or, more usually, unfavourable attitude towards an object, person, thing or situation. It tends to be emotionally charged, stereotyped and not easily changed by information to the contrary. On this basis, people learn to react to a particular category or stimulus in a particular way, which is based on their experiences of similar situations. In some cases, people may be put into the position of categorizing a specific item where there has been no previous experience and in the absence of adequate information. This process of pre-judging is a common aspect of human behaviour.

Perceptual sensitization and defence

These two mechanisms are significant when considering practical applications of perception theory and their contribution to accidents at work.

Perceptual sensitization

People can become 'sensitized' to certain stimuli if they are, in their opinion, relevant or meaningful to them. Thus, stimuli that are relevant to a person are perceived as larger, brighter, more valuable, more attractive than other stimuli. Sensitization is both a permanent and temporary phenomenon. Moreover, it may or may not be accepted consciously by the individual.

Perceptual defence

This mechanism modifies, distorts or eliminates those stimuli that are stressful, threatening or create anxiety. It tends to protect the individual from threatening experiences and situations by obscuring these threats and making them difficult to perceive at a conscious level. Essentially the individual pushes these unwanted intrusions to a subconscious level.

Perceptual sensitization and defence are important in the perception of hazards and risks at work, and may feature in the 'human error' element of an accident.

The perception of hazards

Both perceptual sensitization and defence are important in terms of how people identify hazards. Other factors can have a direct or indirect effect on hazard perception, however. For instance:

- a warning may not be strong enough to penetrate the perceptual set or may be masked by other signals
- an individual who is concentrating heavily on a particular task may not perceive danger that is peripheral or secondary to the task currently being carried out
- some practices or habits, when taken from one situation to another, may create risks, such as driving on motorways compared with driving around a small housing estate
- routine, repetitive, boring and relatively low-skilled tasks produce a tendency to reduce attention to the task, resulting in day-dreaming, which can increase accident potential
- an individual can get used to a certain stimulus, such as a light flashing, and if the significance of this stimulus is not reinforced, it ceases to command attention and, eventually, can be ignored – the process of *habituation*.

Fundamentally this means that:

- people do not see what is *there*, but see what they *expect* to see
- people do not see what they do not *expect* to be present
- people do not see what they do not *want* to be there.

All the above factors directly affect the person's perception of risk and the action to be taken to avert the risk. Perception may also vary according to age, state of health and individual personality differences.

Perceptual illusions

An illusion is a false perception. In other words, what is perceived is not what is physically before the viewer. The apparent size, relative position, distance and relative motion give cues on which mental judgements are made. Perceptual illusions are commonly a sole or contributory cause of accidents.

Changes from the accustomed, that is, what people expect to perceive, result in perceptual errors or visions. Common sources of perceptual error are:

- *size and shape* people judge an object's distance by comparing how large it appears to how large or small they *knew* it was

- *linear perspective* a road appears to reduce in size in the distance – parallel lines converge as they recede and, from this, people mentally calculate the distance to the end of the lines, so variations in width distort the relationship between perceived and actual distance

- *clarity of details* we judge distance by clarity and definition of objects, so an illusion of closeness is created by sharp details being picked out by bright light, while darkness, haze and smoke or dim lighting in a workplace give an illusion of distance

- *relative motion* the object that moves across our visual field rapidly is judged as being closer and vice versa, so a large, close object moving past a person gives the person an illusion of motion even though they are standing still

- *light and shade* many cues of size and distance are based on judgement from shadows, so, for instance, at night, in rain, fog or snow, these cues are not available and distances can be perceived incorrectly

- *flickering illumination* a flickering light can cause false perception of motion and speed as well as causing disorientation and loss of balance

- *autokinesis* this is the phenomenon whereby a single point of light, when stared at, appears to move about, giving a false perception of relative motion – a light on a control, machine or vehicle can have a similar effect

- *refraction* water or glass changes the refraction or deflection of light rays – curved glass panels have the same effect

- *hypoxia* lack of oxygen to the brain can affect vision

- *colour contrast* bright colours move towards the viewer and dim colours recede, so lack of colour contrast reduces the ability to perceive distances between objects, as in the case of camouflage.

Perception and human behaviour

Perception is one of the most important aspects of human behaviour. 'Behaviour' is defined as 'the total response a person makes to any situation with which they are faced'. These responses may be overt or visible, such as words, facial expressions and hand gestures, or convert, such as unvoiced thoughts.

Failure to perceive or faulty perception are, perhaps, two of the greatest causes of human failure and frequently result in accidents at work. In certain cases, the perceptual ability of individuals may need to be measured in order to ensure safe working.

PERSONALITY

Personality is frequently related to the way people behave, for example, rigid, honest, overbearing, flexible and so on. It involves the total pattern of behaviour that is unique and manifest in that person's values, beliefs, interests, attitudes, expressions and actions.

4

Personality involves the traits, ways of adjustment, defence mechanisms and ways of behaving that characterize an individual and their relationship to other people, events and situations.

Allport (1961) defined personality as 'the dynamic organization within the individual of those psychophysical systems that determine his characteristic behaviour and thought'.

The term 'dynamic' implies that personality is composed of interacting parts and that this interaction produces flexibility of response, that is, an individual who is subject to change. This degree of subjection to change is important, particularly in the selection of people who may be exposed to continuing forms of danger. The word 'psychophysical' means that personality contains both mental and physical elements. 'Determine' refers to the notion that personality is considered to be a cause of behaviour.

Personality has also been defined as 'the distinctive characteristics of individuals, the stable and changing relationships between these characteristics, the origins of the characteristics, the ways in which they help or hinder the interaction of a person with other people, and the characteristic ways in which a person thinks about himself' (Morgan and King, 1956).

The structure of personality

Personality fundamentally comprises a series of traits. A trait is a more or less stable and consistent disposition of an individual to respond to people and situations in a characteristic way – friendly, hostile, gregarious, reserved and so on. As constructs or ideas that bring order to the many facets of personality, traits provide what is probably the most useful means of characterizing a person.

Personality is directly connected with:

- abilities, attitudes and interests
- motives
- modes of adjustment
- defence mechanisms.

Adjustment refers to the process of accommodating oneself to circumstances and, more particularly, to the process of satisfying needs, or motives, under various circumstances. *Defence mechanisms* are important adjustment techniques. A defence mechanism is a device, a mode of behaviour, that a person uses unconsciously and automatically to protect themselves against fear, anxiety or feelings of worthlessness that are the emotional consequences of motive frustration. Everyone has their own distinct pattern of defense mechanism and, as such, they are important traits of personality.

Personality factors

The study of personality has produced a wide range of conflicting theories. Probably the best-known approach to personality is based on work by R. B. Catell, who considered responses to questionnaires from people on their various beliefs and preferences. From this research, Cattell produced a list of 16 *personality factors* (see Table 4.1). These factors are shown as 16 dimensions for which a person's level in each case can be recorded to produce a *personality profile*, which is unique to the individual completing the questionnaire.

While any clear link between individual personality and safety is hard to identify, certain personality traits could well be a contributory feature in accidents. For people working in high-risk activities, therefore, there may be a case for the use of the 16 personality factor test.

TABLE 4.1 Cattel's 16 personality factors

◄──────── **1 to 10** ────────►

Low score	High score
Description	*Description*
Reserved, detached, critical, aloof	Outgoing, warmhearted, easy-going, participating
Less intelligent, concrete-thinking	More intelligent, abstract-thinking, bright
Affected by feelings, emotionally less stable, easily upset	Emotionally stable, faces reality, calm, mature
Humble, mild, accommodating, conforming	Assertive, aggressive, stubborn competitive
Sober, prudent, serious, taciturn	Happy-go-lucky, impulsively lively, gay, enthusiastic
Expedient, disregards rules, feels few obligations	Conscientious, persevering, staid, moralistic
Shy, restrained, timid, threat-sensitive	Venturesome, socially bold, uninhibited, spontaneous
Tough-minded, self-reliant, no-nonsense	Tender-minded, clinging, over-protected, sensitive
Trusting, adaptable, free of jealousy, easy to get along	Suspicious, self-opinionated, hard to fool
Practical, careful, conventional, regulated by external realities, proper	Imaginative, wrapped up in inner urgencies, careless of practical matters, Bohemian
Forthright, natural, artless, unpretentious	Shrewd, calculating, worldly, penetrating
Self-assured, confident, serene	Apprehensive, self-reproaching, worrying, troubled
Conservative, respecting established ideas, tolerant of traditional difficulties	Experimenting, liberal, analytical, free-thinking
Group-dependent, a 'joiner' and sound follower	Self-sufficient, prefers own decisions, resourceful
Undisciplined self conflict, follows own urges, careless of protocol	Controlled, socially precise, following self-image
Relaxed, tranquil, unfrustrated	Tense, frustrated, driven, overwrought

4

MEMORY

Memory is the process of retaining, recognizing and recalling experience (remembering) and is particularly associated with how people learn things. It is, essentially, the storage facility of the human system. We have both *short-term* and *long-term memory*.

Short-term memory

Short-term or immediate memory refers to the temporary storage of information for a few seconds, as in the case of a telephone number. It is also associated with the amount of information one can take in and retain and is, in many cases, a limiting factor on individual ability and safety. The limited storage capacity of short-term memory is also shown by the fact that the memory span for a single repetition is about seven items long. Without regrouping or reorganizing the information as we receive it, most people cannot retain more than about seven items, for example, numbers, after one exposure to them. This fact has been demonstrated in experiments with individual subjects memorizing numbers, the ability to retain these numbers reducing as the number of digits increases. Short-term memory is, thus, limited in capacity and highly susceptible to disruption or interference as compared with long-term memory.

This said, people have learned to recode the information into chunks or bits, and the trick is to recode the information as it is received. Most people can only, however, deal with around 3.1 bits of information at any point in time. Once this level is exceeded, the ability to recall the information reduces proportionately. This fact is aptly demonstrated when seeking directions from other people. In many cases, the person giving the directions provides so much information in an attempt to help that, by the time they have finished, the first part of the instructions are confused or lost!

Short-term memory and its limitations can be a significant factor in the causes of accidents at work, and one that is frequently associated with human error or poor memory skills. Important on-the-spot instructions should, therefore, be repeated several times to ensure that the recipient fully understands and can recall them.

Long-term memory

Long-term memory is concerned with the ability to store and subsequently recall information. It is a vast store of information that is organized in some form of classification. On this basis, any new information is perceived in terms of these categories and forced into the classification system even when it does not fit exactly. In this process there is a chance that it may become distorted.

Long-term memory is developed from an early age through the repetition of items and codifying them to produce a meaning. (A pnemonic is an example of a codifying system.) There is a characteristic drop in memory over a period of time associated with the ageing process.

Interference with long-term memory can be caused by:

- events of close similarity that tend to confuse
- the effect of recall on the subsequent memory that can, again, cause confusion

resulting in the individual forgetting.

Limitations in memory recall, or remembering, can frequently be overcome by recalling the circumstances in which the original memory was stored or approaching it via memories that we know were associated with it. Unavailability of memory may be significant in certain emergency situations or where a quick response is required from an individual. This unavailability can be overcome by recalling and reusing the memories (knowledge and skills) at regular intervals. Various forms of refresher training, fire and emergency drills and practical sessions all assist in reducing unavailability of memory.

Variables that influence the amount retained in long-term memory are:

- the meaningfulness of the material
- the degree of learning of the material
- interference with the learning material.

In many cases, memory may undergo significant change over a long period of time. This is particularly common where a memory may be unpleasant and it is likely that distortion will occur at each recall. Thus, these various recalls, distorted to various degrees, will eventually be remembered rather than the original version. A stage is reached where the individual is unable to distinguish between the correct facts and those that were introduced as part of subsequent recalls. Such a phenomenon is commonly encountered in accident investigation whereby a witness may make a particular statement following the accident, but, three months later, after they have endeavoured to recall the situation on numerous occasions, may have a totally different version of the events leading to the accident.

Memory defects are sometimes associated with the phenomenon of *accident proneness* – the situation where a person has the same accident time and time again (see later in this chapter).

CONCLUSION

Memory is an important feature of human behaviour, particularly where people may be exposed to hazards. It is directly affected by learning, past experience, feedback from events of particular significance and an individual's capacity for storing information.

INFORMATION PROCESSING

For an individual to act in a certain situation, it is necessary for them to process the information presented at that point in time. Thus, the relative speed with which people process information is a pre-condition of many accidents.

Each stimulus produces a reaction or response and the response can be divided into two elements:

- *specific reaction time* the actual time it takes to perceive and process the response
- *movement time* the time taken to actually execute the response.

Should a second stimulus arrive during the movement time, then it has to wait until the first stimulus has been dealt with. The actual movement in response is being *monitored*, the single channel process ensuring that the execution of the original response was accurate. The particularly crucial parts of the movement are the beginning and the end, the middle part often being partially or totally neglected.

Generally, people cannot do more than one thing at a time, the speed and sequence of response varying from person to person. This factor can be significant in accident causation.

With well-known and practised tasks, such as operating a machine or driving a vehicle, the monitoring action of the brain can be reduced, depending on the speed with which a person can respond to stimuli and not monitor specific movements or actions. Results are achieved through continuous practise or the speed–accuracy trade-off, whereby the monitoring is voluntarily removed.

The *feedback* that people receive is an important feature of monitoring a task. Where an individual may be highly skilled, it can be a hindrance and actually destroy performance. In the teaching, for instance, of learner drivers or certain activities, such as golf, the instructor has to put the monitoring aspect back into the task. This can adversely affect the individual's level of performance.

Information processing can be separated into *on-line processing* and *off-line processing*.

4

On-line processing

This is the spur of the moment decision-making that an individual has to take in order to survive and is associated with the single channel theory above. On-line processing is of limited capacity and can best be used by grouping actions together as habits or packages that can be put into operation as a group sooner than they could as a series of separate actions.

This process is commonly associated with jobs that incorporate repetitive skills, such as feeding components into a machine, certain features of driving vehicles, such as starting, stopping and changing gear. These habits, which are developed over a lifetime, can be difficult to break because of the fact that the brain's monitoring element is greatly reduced or completely absent. This fact can be a contributory feature in accidents, particularly where the machine develops a fault and the operator continues the process without realizing that the fault has occurred.

Induction and specific training programmes endeavour to inculcate the *correct* habits, including dealing with emergencies, right from the start, with a view to preventing human error-related accidents.

Off-line processing

This is a process whereby people actually simulate in their minds the outcomes of different courses of action prior to making any final decision as to which course of action to take. This planning process, and the individual skills that people possess for such planning, is an important feature in predicting accidents or other adverse consequences that could arise.

Off-line processing is a skill based on individual knowledge, intelligence, experience of similar situations and the amount of practice in using this skill. It has its limitations, because, for example there are attempts to simplify decisions, perhaps by ignoring less important aspects, in order to make it more easily manageable in the mind. These limitations, some of which are unconsciously brought in, can result in incorrect or unsafe decisions being made by an operator.

In the information processing operation, people ascribe different values to various outcomes of their decisions. These values may be influenced by extraneous factors, such as any financial benefits to be derived or the possibility of saving time and effort, which are subjective and influenced by the personality of the person making the decision and previous experience of similar situations.

The level of brain arousal can also affect the efficiency and rate of mental processing. *Arousal* is defined as an increase in alertness and muscular tension. Levels of arousal vary significantly. Generally, at low arousal levels, performance is poor. As arousal increases to an optimum, performance rises accordingly, but then drops as further arousal takes place. Changes in arousal levels take place during the average working cycle.

Capacity to act

Capacity to act following a decision will vary according to the physical and mental limitations of people, such as their strength, speed and so on. No two people, therefore, react to a particular stimulus in the same way or at the same speed. Speed of reaction may be particularly important in the case of vehicle accidents.

Skills and accidents

Basic skills have a direct relationship with accidents and their causes. Factors affecting individual skills in relation to accidents include the following.

- *Reaction time* Simple speed of reaction has little significance as a causative factor in accidents. Choice reaction, where there are several stimuli and several available responses, however, is significant. A typical example is where an operator may need to select a specific control from several available controls on a machine to avoid a dangerous situation.

- *Coordination* Manipulation and dexterity tests, such as the running of a loop around a distorted electric wire, and which require a considerable degree of hand and eye coordination, can be used to predict accident potential in certain cases, such as driving activities.

- *Attention* Where attention is divided between several stimuli, and the operator must respond to all these stimuli at the same time, a considerable degree of skill is required. Attention, or lack of it, is an important feature of many accidents.

Many of the tests used in the selection of personnel are based on the factors of reaction time, coordination and attention. However, the correlation between skills and accidents is only partial. Generally, the more closely the test parallells the task to be carried out, the more accurate will be any prediction or assessment as to future accident potential.

CONCLUSION

How people process information in a variety of situations is an important aspect of human performance, particularly where they may be exposed to danger. No two people process information in the same way or at the same rate.

Many factors are associated with the human information processing system – the level of arousal, personality, individual feedback from previous situations, intelligence and personal monitoring ability.

ACCIDENT PRONENESS

The concept of accident proneness , that is, that some people are more susceptible to accidents than others, is central to the psychology of accidents. Accident proneness is a truism, but true accident proneness exists in an infinitesimally small number of cases, and not to such an extent that it warrants consideration as a causative factor in accidents.

Research by A. R. Hale and M. Hale established that both age and experience were correlated with differences in accident susceptibility.

Accident proneness has, in the past, been linked with the specific personality traits or characteristics of individuals. It has also been linked with intelligence, coordination and attention skills, stressful periods in people's lives, and the notion that certain individuals may go through periods of accident proneness that tend to be of a temporary nature.

CONCLUSION

The concept of accident proneness, or the notion that some people have a greater susceptibility to accidents than others, has been the subject of research over many years. Generally, the outcome of this research is inconclusive and, on this basis, accident proneness does not rank as having any significant importance as a causative factor in accidents.

RISK TAKING

Risk taking is a standard feature of human behaviour, but what do we mean by 'risk'? Various definitions of the term can be noted – 'a chance of loss or injury', 'the probability of harm, damage or injury', 'to expose to mischance' and 'the probability of a hazard leading to personal injury and the severity of that injury'.

There is no doubt that people take risks in all sorts of situations and for a variety of reasons, such as financial gain or to save time and effort. Some of the more well-known theories relating to risk taking are outlined below.

Mathematical models of people's behaviour

The American school of psychologists said that one can consider risk taking to be an attempt by the individual to maximize some function of probability and the value of the outcome of their decision. The logical way is to calculate probability or likelihood and the value of all the possible outcomes and rank them according to degree. Experiments have involved mainly gambling situations, where it is possible to relate the probability of winning to the financial value of taking the risk. However, this simple equation does not predict people's behaviour and is crude from a statistical viewpoint. For instance, it does not explain why people will gamble on situations where there is a very low probability of winning, such as the football pools.

Other factors affecting risk and the value of the outcome are the urge to win and the severity of loss or injury resulting from the risk-taking activity. In gambling situations, individual differences between the sorts of risk people prefer to take have been noted. A study of compulsive gamblers showed that the probability of winning was the overriding factor in their taking the risks, the value of the outcome being less significant. In other words, the greater the risk, the happier they were, the situation being seen as a 'people battle' between the gambler and fate.

Risk taking in relation to arousal

Arousal is defined as an increase in alertness and muscular tension. It is an emotion that may be associated with fear, expectation, excitement, physical exercise and certain stressful events. It is characterized by a wide range of symptoms – pounding heart and rapid pulse, tense muscles, dryness of the throat and mouth, trembling, nervous perspiration (cold sweat) or a feeling of sickness – and may be artificially induced by using drugs.

Most people endeavour to actually avoid taking risks as far as possible. There are some people, however, who enjoy various forms of risk taking, such as compulsive gamblers. The relationship between arousal and the level of personal risk taking is based on the fact that when people take risks – physical risks, such as driving too fast, financial risks, such as playing the stock market, or social risks, for instance indulging in behaviour that could result in loss of face – it increases their level of arousal to an optimum level or 'high'. Once this level of arousal has been reached, they then tend to take fewer risks.

Most studies have involved driving situations. F. W. Taylor examined the aspect of galvanic skin response, the basis of the lie detector test, in driving experiments and found a relationship between road conditions and the level of arousal of drivers. In other words, the worse the conditions, the higher the level of arousal that was maintained by the drivers concerned.

Risk thresholds

Attempts have been made to measure people's risk thresholds, that is, the levels that they would prefer to operate at in terms of risk. They considered physical risk taking, such as walking in front of a bus, financial risk taking, for example, insurance-related activities, playing the stock market, gambling on horse racing, and social risk taking.

While the results are not clear-cut, a broad threshold across these areas emerges for some people, which is combined with a significant personality factor or variant, the latter being the more important.

Achievement motivation

J. W Atkinson related risk taking to the concept of *achievement motivation*, which is derived from motivation theory. He showed that there are various factors of motivation affecting people, in particular fear of failure (negative) and motivation towards success (positive). Such factors can be measured by a well-validated questionnaire. These views stemmed from a larger study of business risk taking by D. C. McClelland. He examined the factors that motivated people to establish their own businesses, which clearly involves some degree of financial risk, as opposed to the relative job security that comes from being employed by a large organization. He showed that achievement motivation, the desire to 'do your own thing', was a strong motivator for the first-mentioned group, that they actually enjoyed the financial risk taking involved.

Skill versus chance

There are great differences in people's behaviour in the skill versus chance situation. With skilled tasks, for instance, people may find the risk challenging. They feel their behaviour, and the skills they possess, are influencing the situation. However, when people perceive that their behaviour does not affect the situation, that it is a chance situation, then their behaviour is largely determined by the value of the outcome and the probability factor. In other words, they act more directly towards the mathematical model. However, when skill enters the situation, the value of the risk is significant – a challenge to the individual's behaviour.

The skill versus chance situation is important in the evaluation of occupational risks, as follows:

- *value of outcomes* affected by:
- punishment
- financial gain
- the severity of harm or injury
- other people's opinion

- *probability – level of risk affected by*
- personality
- the situation.

Individual risk taking is, thus, associated with the probability of harm arising and the perceived value of the outcome for the risk taker. However, individual judgements or assessments of risk varies considerably. R. Robaye, *et al.*, got people to assess the risk in various work situations by examining photographs and then correlated the risks with actual accident rates. They established, for instance, that people who have a lot of accidents tend to judge the risks higher and vice versa, but they judged the consequences of taking the risks as being much less serious, that is high probability, low severity.

CONCLUSION

Accidents can be caused for a number of reasons:

- **the information received by the individual is incorrect, namely the objective danger is greater than the subjective perception of the risk involved**

- the individual does not possess the necessary skills, both physical and intellectual, which is particularly common where the increase in danger is so small as to be imperceptible
- the individual's motivation is inappropriate to the circumstances.
- the probability matrix is incorrect
- feedback from previous situations is ineffective, that is, people are not learning from their past experiences.

In endeavouring to bring about changes in behaviour, namely attitudes to safe working, an assessment as to which of the above five factors is significant must be made. Inappropriate motivation and incorrect assessment of probability are the factors that most need to be considered in the promotion of safe behaviour.

THE POTENTIAL FOR HUMAN ERROR

The study of human factors is very much concerned with identifying those aspects of behaviour that result in people making mistakes or errors, some of which could result in accidents or incidents of various kinds. Limitations in human capacity to perceive, attend to, remember, process and act on information are all relevant in the context of human error.

The HSE's publication, 'Human factors and industrial safety' (HS(G)48, 1989), identifies a number of factors that can contribute to human error and the resulting accidents. These include the following.

- *Inadequate information* People do not make errors merely because they are careless or inattentive. Often they have understandable (albeit incorrect) reasons for acting in the way they did. One common reason is ignorance of the production processes in which they are involved and of the potential consequences of their actions.
- *Lack of understanding* This often arises as a result of a failure to communicate accurately and fully the stages of a process that an item has been through. As a result, people make presumptions that certain actions have been taken when this is not the case.
- *Inadequate design* Designers of plant, processes or systems of work must always take into account human fallibility and never pre-

sume that those who operate or maintain plant or systems have a full and continuous appreciation of their essential features. Indeed, failure to consider such matters is, itself, an aspect of human error.

Where it cannot be entirely eliminated, error must be made evident or difficult. Compliance with safety precautions must be made easy. Adequate information as to hazards must be provided. Systems should 'fail safe', that is, refuse to produce in unsafe modes of operation.

- *Lapses of attention* The individual's intentions and objectives are correct and the proper course of action is selected, but a slip occurs in performing it. This may be due to competing demands for (limited) attention. Paradoxically, highly skilled performers, because they depend on finely tuned allocation of their attention, to avoid having to think carefully about every minor detail, may be more likely to make a slip.

- *Mistaken actions* This is the classic situation of doing the wrong thing under the impression that they are right. For example, the individual knows what needs to be done, but chooses an inappropriate method to achieve it.

- *Misperceptions* Misperceptions tend to occur when an individual's limited capacity to give attention to competing information under stress produces `tunnel vision' or when a preconceived diagnosis blocks out sources of inconsistent information. There is a strong tendency to assume that an established pattern holds good so long as most of the indications are to that effect, even if there is an unexpected indication to the contrary. One potent source of error in such situations is an inability to analyse and reconcile conflicting evidence deriving from an imperfect understanding of the process itself or of the meaning conveyed by instruments. Full analysis of the preventative measures required involves the need for people to understand the process as well as technical and ergonomic considerations concerned with instrumentation.

The official report on the accident in 1979 at the Three Mile Island nuclear power station in the USA, cited human factors issues as the main causes. Misleading and badly presented operating procedures, poor control room design, inadequate training and poorly designed display systems all, in one way or another, gave the operators misleading or incomplete information. In the event, the radioactive exposure off the site was very small indeed. However, the official enquiry emphasized how failures in human factors design, inadequate training and procedures, and inadequate management organization led to a series of relatively minor technical faults being magnified into a near disaster with significant economic and possible human consequences.

- *Mistaken priorities* An organization's objectives, particularly the relative priorities of different goals, may not be clearly conveyed to, or understood by, individuals. A crucial area of potential conflict is between safety and other objectives, such as output or the saving of cost or time. Misperceptions (described above) may then be partly intentional as certain events are ignored in the pursuit of competing objectives. When top management's goals are not clear, individuals at any level in the organization may superimpose their own.

- *Wilfulness* Wilfully disregarding safety rules is rarely a *primary* cause of incidents. Sometimes, however, there is only a fine line between mistaken priorities and wilfulness. Managers need to be alert to the influences that, in combination, persuade staff to take (and condone others taking) short cuts through the safety rules and procedures because, mistakenly, the perceived benefits outweigh the risks, and they have perhaps got away with it in the past.

THE SIGNIFICANCE OF PERSONAL FACTORS

Employees bring personal habits, attitudes, skills, personality and other factors to their jobs that, in relation to task demands, may be strengths or weaknesses. Individual characteristics influence behaviour in complicated and significant ways. Their effects on task performance may be negative and cannot always be mitigated by job design solutions. Some characteristics, such as personality, are fixed and largely incapable of modification. Others, such as skills and attitudes, are amenable to modification or enhancement. The person, therefore, needs to be matched to the job.

Important considerations within the personal factors category include the following.

- *Task analysis* Thorough task analysis, especially for critical jobs, a detailed job description should be generated. From this, a specification can be drawn up to include such factors as age, physique, skills, qualifications and experience, aptitude, knowledge, intelligence and personality. Personnel selection policies and procedures should ensure that the specifications are matched by the individuals.

- *Training* Training will produce an employee capable of working without close supervision, confident to take on responsibility and perform effectively, providing that initial selection is carried out with care. Training, both induction and continuation, benefits individuals themselves as well as their colleagues. Self-confidence and job satisfaction grow significantly when people are trained to work correctly under both routine and emergency conditions. Training should aim to give all individuals the skills to allow them to understand the workings of the plant and processes. It is not a once-and-for-all activity, for when procedures and processes change and complicated skills, particularly when underused, deteriorate these influence an individual's performance.

4

- *Monitoring of performance* Monitoring of personal performance, apart from providing appropriate feedback on such performance, includes supervision of safety practices and other approaches for developing an effective climate of opinion towards safety.

- *Fitness for work and health surveillance* For certain jobs, there may be specified medical standards for which pre-employment and/or periodic health surveillance is necessary. These may relate to the functional requirements of the job or the impact of specified conditions on the ability to perform it adequately and safely. An example is the medical examination of divers. There may also be a need for routine health surveillance for the effects of exposure to workplace hazards, both physical, such as the effects of acute heat stress, or chemical, for example, absorption of organo phosphorous insecticides, which may impair operational ability.

  A review of health on return to work following a period of sickness absence, the recognition of the place of counselling and advice during periods of need or stress and possible need for redeployment should also be considered. In this context, alcohol or drug abuse, and the possible adverse side-effects of prescribed drugs, are also relevant. Access to specialist assistance may be necessary to deal with these aspects.

CONCLUSION

Individual aspects of human behaviour are affected by many factors. With the increased emphasis on safe person strategies in safety and accident prevention, it is well to consider many of the aspects of human behaviour discussed in this chapter when looking at the area of human capability as part of the risk assessment process.

5

Accident causation

INTRODUCTION

Why do accidents happen? This is a familiar question and one that has resulted in much research since the turn of the century. Can the cause of accidents be such that a general pattern of the accident phenomenon will emerge? To put it simply, a theory of the accident causation process can be written if most of the components of the accident process can be discerned and if the factors that contribute and lead to the accident can be discovered.

THEORIES OF ACCIDENT CAUSATION

The pure chance theory

This theory is that everyone in the population has an equal chance of having an accident. It suggests that no discernible pattern emerges in the events that lead up to an accident. An accident is usually treated as an act of God, leaving one to accept the fact that prevention is non-existent.

The biased liability theory

This theory proposes the idea that once a person has an accident, the probability that the same person will have a further accident in the future has either decreased or increased when compared to the rest of the population at risk. If the probability has increased, the phenomenon is referred to as the *Contagion Hypothesis*. If the probability has decreased, it is commonly called the *Burned Fingers Hypothesis*.

The accident proneness (unequal initial liability) theory

This has been the most widely discussed theory in the history of accident research. It proposes that there exists a certain subgroup within the general population that is more liable to incur accidents. This theory refers to some innate personality characteristics that cause accident prone individuals to have more accidents than non-accident prone people.

The theory of unconscious motivation

This has its roots in psychoanalytic theory. The idea is that accidents are brought about by subconscious processes, including guilt, aggression, anxiety, ambition and conflict. The theory focuses only on the individual and the interaction of their perception of the environment with their underlying personality factors.

The adjustment – stress and goals – freedom – alertness theories

These are two complementary theories, developed by J. Kerr in 1950 and 1957. The first theory states that individuals who fail to reach some sort of adjustment with their working environment will tend to have more accidents than others. This adjustment is affected by physical and psychological stressors.

The second theory postulates that individuals have accidents due to a lack of alertness brought about by the fact that such people have no freedom in choosing the goals set for their working situation. Kerr (1957) stated that freedom to set goals results in high-quality work performance. The level of quality should rise as the level of alertness increases. In this context, if the working climate is made more rewarding, and if the individual feels that they have some degree of control over their working environment, then the level of alertness will increase. In contrast, increased stress on an individual in a working situation will increase the probability that accidents will occur.

The Domino Theory

One of the more colourful theories of accident causation was formulated by H. W. Heinrich in 1959, and is known as the Domino Theory. This theory explains the accident process in terms of five factors:

- ancestry and social environment
- fault of the person
- the unsafe act and/or mechanical or physical hazard
- the accident
- the injury

These factors are of a fixed and logical order. Each one is dependent on the one immediately preceding it, so that if one is absent, no injury can occur. The theory can be visualized as five standing dominoes and the behaviour of these dominoes is studied when they are subjected to a disturbing force. When the first, social environment, falls, the other four automatically follow, unless one of the factors has been corrected, that is, removed, thereby creating a gap in the required sequence for producing an accident.

5

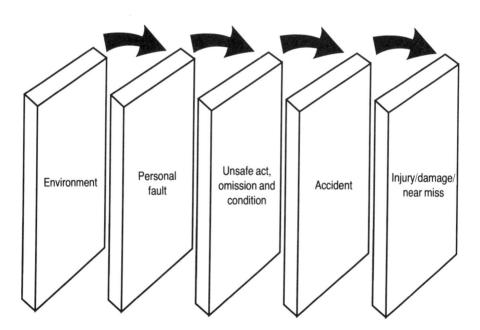

- **FIG 5.1 Heinrich's Domino Theory**

The Epidemiological Approach

This approach arose as a result of the formalized study of epidemics, and is more a description of the accident environment than it is an analytical theory. It is used to provide a conceptual framework for many complicated factors that affect the human situation and in an increased incidence of accidents. According to J. Haddon, et al. (1964), the accident is caused by the conjunction of the following:

- *host factors* these are related to the person who sustained the accident and include such aspects as the age and sex of the victim
- *agent* this is the object that directly gave rise to the accident and can be considered as either the types of abnormal energy exchanges that produced the injury or the specific types of damage produced
- *environment* this is further subdivided into:
- physical, such as geography, seasons and so on
- biological, for example, poisons, toxins
- socio-economical – some individuals are more susceptible to accidents than others.

The modelling approach

The modelling approach in safety research has been recommended by several authors (for example, A. R. Hale and M. Hale, 1970, who produced a simulation of the accident process). This approach allows the investigator to graphically depict the accident phenomenon.

In Hale and Hale's model (see Figure 5.2), use is made of a closed-loop system, which considers the major factors of:

- presented, expected and perceived information
- the action
- the feedback.

Secondary features, such as increasing age, past experience and so on are also considered. The typical situation involves *presented information*, which may be incorrect or incomplete, and can be affected by such factors as physical problems of the individual and the design/layout of the workplace. The *expected information* is a function of past experiences and population stereotypes. Presented and expected information combine into *perceived information*, which is affected by such

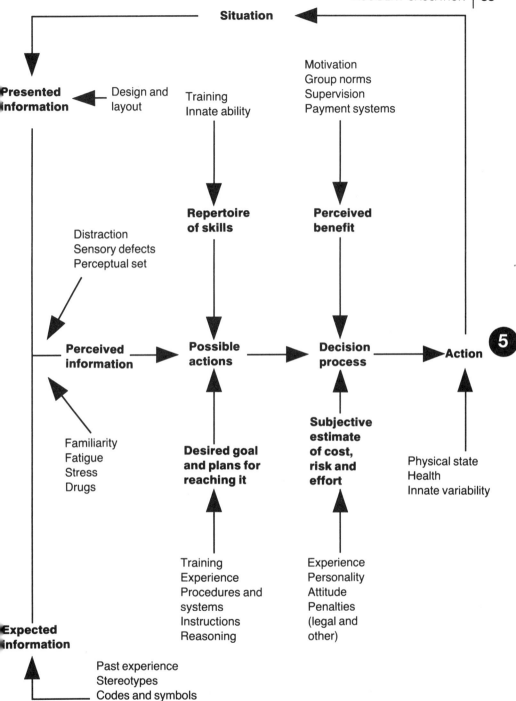

● FIG 5.2 Model of human performance in relation to accident causation
(A. R. Hale and M. Hale, 1970)

secondary factors as the effects of fatigue or drugs, sensory defects and distractions. *Possible actions* are then listed according to the individual's repertoire of skills, desired goal and plans. The individual then goes through the *decision-making process*, which is based on the perceived benefit and a subjective estimate of the cost, risk and effort in taking the considered action. Finally, there is the *action* that could result in the accident, again affected by the individual's state of · health, physical state and various innate variabilities. This action results in feedback to the next situation.

An updated Domino sequence

F. E. Bird and R. G. Loftus have expanded on the Domino Theory to reflect the influence of management in the cause and effect of all accidents that result in a wastage of the organization's assets. The modified sequence of events becomes:

- lack of control by management, permitting
- basic causes (personal and job factors), that lead to
- immediate causes (substandard practices/conditions/errors), which are the proximate causes of
- the accident, which results in
- the loss (minor, serious or catastrophic).

This modified sequence can be applied to all accidents, and is fundamental to loss control management (see Figure 5.3).

The International Loss Control Institute's loss causation model

The International Loss Control Institute's model takes into account all potential loss situations. This model shows that all potential losses associated with accidents, property damage, occupational ill-health and so on are due to lack of control on the part of management, namely an inadequate loss control programme, inadequate programme standards and/or inadequate compliance with established standards. This lack of control leads to the basic causes of accidents – personal and job factors. Such factors are the immediate causes leading to the defined incident. The outcome of this chain of events is some form of loss to people, property and/or process (see Figure 5.4).

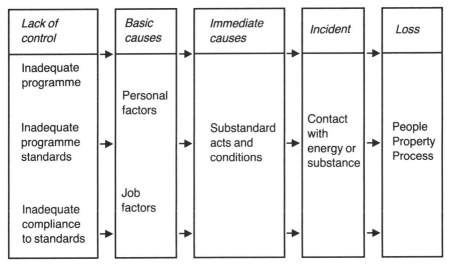

- **FIG 5.3 Updated Domino Sequence (Bird and Loftus)**

Multiple causation theory

Multicausality means that there may be more than one cause of an accident, as shown in Figure 5.5.

Each of these multicauses is equivalent to the third domino in Heinrich's theory and can represent an unsafe act, condition or situation. Each of these can itself have multicauses and the process during accident investigation of following each branch back to its root is known as *fault tree analysis*.

The theory of multicausation is that the contribution causes combine together in a random fashion to result in an accident. During accident investigations, there is a need to identify as many of these causes as possible.

The cause–accident–result sequence

This theory proposes that the indirect causes (personal factors and source causes) contribute to the direct causes (unsafe acts and unsafe conditions) that result in an accident. Similarly, the accident has direct and indirect results for the injured person and the organization (see Figure 5.6).

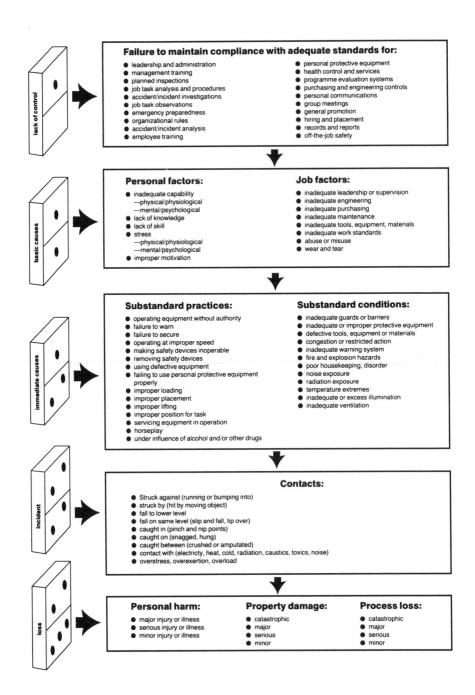

Failure to maintain compliance with adequate standards for:

- leadership and administration
- management training
- planned inspections
- job task analysis and procedures
- accident/incident investigations
- job task observations
- emergency preparedness
- organizational rules
- accident/incident analysis
- employee training
- personal protective equipment
- health control and services
- programme evaluation systems
- purchasing and engineering controls
- personal communications
- group meetings
- general promotion
- hiring and placement
- records and reports
- off-the-job safety

lack of control

Personal factors:

- inadequate capability
  —physical/physiological
  —mental/psychological
- lack of knowledge
- lack of skill
- stress
  —physical/physiological
  —mental/psychological
- improper motivation

Job factors:

- inadequate leadership or supervision
- inadequate engineering
- inadequate purchasing
- inadequate maintenance
- inadequate tools, equipment, materials
- inadequate work standards
- abuse or misuse
- wear and tear

basic causes

Substandard practices:

- operating equipment without authority
- failure to warn
- failure to secure
- operating at improper speed
- making safety devices inoperable
- removing safety devices
- using defective equipment
- failing to use personal protective equipment properly
- improper loading
- improper placement
- improper lifting
- improper position for task
- servicing equipment in operation
- horseplay
- under influence of alcohol and/or other drugs

Substandard conditions:

- inadequate guards or barriers
- inadequate or improper protective equipment
- defective tools, equipment or materials
- congestion or restricted action
- inadequate warning system
- fire and explosion hazards
- poor housekeeping, disorder
- noise exposure
- radiation exposure
- temperature extremes
- inadequate or excess illumination
- inadequate ventilation

immediate causes

Contacts:

- Struck against (running or bumping into)
- struck by (hit by moving object)
- fall to lower level
- fall on same level (slip and fall, tip over)
- caught in (pinch and nip points)
- caught on (snagged, hung)
- caught between (crushed or amputated)
- contact with (electricty, heat, cold, radiation, caustics, toxics, noise)
- overstress, overexertion, overload

incident

Personal harm:

- major injury or illness
- serious injury or illness
- minor injury or illness

Property damage:

- catastrophic
- major
- serious
- minor

Process loss:

- catastrophic
- major
- serious
- minor

loss

● **FIG 5.4 The International Loss Control Institute's Loss Causation model in detail**

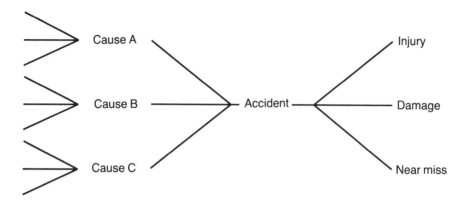

● **FIG 5.5 Multiple causation theory**

CONCLUSION

The causes of accidents are many and varied. There are both indirect and direct causes of accidents, and indirect and direct results of accidents.

A study of the various theories of accident causation indicated the following human factors as being significant, but not necessarily in any order of significance:

- physical and psychological stressors
- environmental stressors
- lack of alertness
- lack of individual control over situations
- unsafe acts and conditions
- lack of management supervision and control
- personal factors, such as attitude, motivation and perception
- information processing defects
- various job-related factors
- the perceived benefits of a particular course of action.

All these factors need to be considered in the study of accident causation and in the investigation of accidents.

INDIRECT CAUSES →	DIRECT CAUSES →	ACCIDENTS →	DIRECT RESULTS →	INDIRECT RESULTS
Personal factors *Definition:* any condition or characteristic of a person that causes or influences them to act unsafely: ● knowledge and skill deficiencies: – lack of hazard awareness – lack of job knowledge – lack of job skill ● conflicting motivations: – saving time and effort – avoiding discomfort – attracting attention – asserting independence – seeking group approval – expressing resentment ● physical and mental incapacities. **Source causes** *Definition:* any circumstance that may cause or contribute to the development of an unsafe condition. *Major sources:* ● production employees ● maintenance employees ● design and engineering ● purchasing practises ● normal wear through use ● abnormal wear and tear ● lack of preventive maintenance ● outside contractors.	**Unsafe acts** *Definition:* any act that deviates from a generally recognized safe way of doing a job and increases the likelihood of an accident. *Basic types:* ● operating without authority ● failure to make secure ● operating at unsafe speed ● failure to warn or signal ● nullifying safety devices ● using defective equipment ● using equipment unsafely ● taking unsafe position ● repairing or servicing moving or energized equipment ● riding hazardous equipment ● horseplay ● failure to use protection. **Unsafe conditions** *Definition:* any environmental condition that may cause or contribute to an accident. *Basic types:* ● inadequate guards and safety devices ● inadequate warning systems ● fire and explosion hazards ● unexpected movement hazards ● poor housekeeping ● protruding hazards ● congestion, close clearance ● hazardous atmospheric conditions ● hazardous placement or storage ● unsafe equipment defects ● inadequate illumination, noise ● hazardous personal attire.	**The accident** *Definition:* an unexpected occurrence that interrupts work and usually takes the form of an abrupt contact. *Basic types:* ● struck by ● contact by ● struck against ● contact with ● caught in ● caught on ● caught between ● fall to different level ● fall on same level ● exposure ● over exertion/strain.	**Direct results** *Definition:* the immediate results of an accident. *Basic types:* ● 'no results' or near miss ● minor injury ● major injury ● property damage.	**Indirect results** *Definition:* the consequences for all concerned that flow from the direct results of accidents. *For the injured:* ● loss of earnings ● disrupted family life ● disrupted personal life ● and other consequences. *For the company:* ● Injury costs ● production loss costs ● property damage costs ● lowered employee morale ● poor reputation ● poor customer relations ● lost supervisor time ● product damage costs.

● **FIG 5.6 The Course–Accident–Result sequence**

Communication

6

INTRODUCTION

To communicate means to impart or transmit. Communication is the transfer of information, ideas, feelings, knowledge and emotions between one individual or group of individuals and another, the basic function of which is to convey meanings.

The objectives or goals of communication are:

- to understand others
- to obtain clear reception or perception of information
- to obtain understanding
- to achieve acceptance (that is, agreement and commitment) of ideas
- to facilitate or obtain effective human behaviour or action.

THE COMMUNICATION PROCESS

Communication involves both a communicator, the originator of the message, and a receiver, the recipient of the message. Communication commonly takes place in four phases or stages, thus:

1 *transmission, or sending out,* of data, both cognitive data and emotional
2 *receiving or perceiving the data*
3 *understanding* the data
4 *acceptance* of the data.

The various stages of the communication process can be shown as in Figure 6.1.

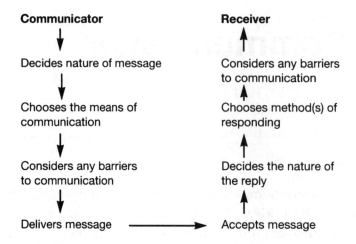

Communicator	Receiver
↓	↑
Decides nature of message	Considers any barriers to communication
↓	↑
Chooses the means of communication	Chooses method(s) of responding
↓	↑
Considers any barriers to communication	Decides the nature of the reply
↓	↑
Delivers message ――――――→	Accepts message

● **FIG 6.1 The stages involved in conscious communication**

Barriers to communication

A number of barriers can arise at the various points during the phases of the process, in particular:

● *barriers to reception* reception of communication can be influenced by:
- the needs, anxieties and expectations of the receiver/listener
- the attitudes and values of the receiver
- environmental stimuli, such as noise

● *barriers to understanding* understanding is a complicated process and is affected by:
- the use of inappropriate language, technical jargon
- the extent to which the listener can concentrate on receiving the data completely, that is variations in listening skills
- prejudgements made by the listener
- the ability of the listener to consider factors that may be disturbing or contrary to their ideas and opinions, that is the degree of open-mindedness that they possess
- the length of the communication
- the degree of knowledge possessed by the listener

- *barriers to acceptance* acceptance of a communication is affected by:
  - the attitudes and values of the listener
  - individual prejudices held by listeners
  - status clashes between the sender and receiver
  - interpersonal emotional conflicts.

COMMUNICATION WITHIN ORGANIZATIONS

Many forms of communication exist within organizations.

- *Formal and informal communication* That which is officially inspired by directors and managers is often referred to as *formal* communication. Communication that is unofficial, unplanned and spontaneous is classified as *informal* communication.

- *One-way and two-way communication* Communication may be *one way*, for example, an instruction to staff from a senior manager, or *two way*, where, following transmission of a message, the views of certain persons are sought prior to the decision-making process.

- *Verbal, non-verbal and written communication* The form of communication is significant. Communication may be by word of mouth (*verbal communication*), by the use of gestures, eye contact and terminal glances (*non-verbal communication*) and by use of memoranda, letters and reports (*written communication*).

- *Intentional or unintentional* Communication may be *intentional*, that is, consciously transmitted, or *unintentional*, where the information is involuntarily transmitted. Significant features of the communication process are, first, the actual lines or channels of communication and, second, the relative effectiveness of the different forms of communication.

FUNCTIONS OF COMMUNICATION

There are many functions of communication. They can be summarized as follows:

- *instrumental* to achieve or obtain something, such as improved safety performance or greater commitment to ensuring safe working practices

- *control* to get someone to behave in a particular way, for example, a person who may not be following a safe system of work

- *information* to establish facts or, alternatively, to explain something to a person

- *expression* to express feelings, such as fear, anxiety, happiness, guilt, or convey oneself in a particular way

- *social contact* to enjoy another person's company

- *alleviation of anxiety* to sort out a problem or ease a worry

- *stimulation* to increase or raise the interst of an individual or group

- *role-related* to give instructions, advice or a warning to a subordinate because the situation requires it.

THE DIRECTION OF COMMUNICATION

One-way communication

One-way communication may be appropriate in certain circumstances and is certainly faster than two-way communication. As such, it does not permit any form of feedback from the receiver of the communication. It is used principally in the giving of directions, instructions and orders.

Two-way communication

This has been found to be far more effective. As such, it gives people the chance to use their intelligence, to contribute knowledge, to participate in the decision-making process, to fulfil their creative needs and to express agreement or disagreement. It helps both the sender and the receiver to measure their standard of achievement and when they both see that they are making progress, their joint commitment to a task will be greater. In some circumstances, the sender may feel that they are under attack as the receiver will identify their errors and inform them of what they are. However, this tends to be more helpful in that a frank discussion will lead to a higher level of understanding and acceptance than one that skirts around the issues.

VERBAL AND NON-VERBAL COMMUNICATION

Verbal communication

Verbal communication should not be treated solely in terms of the verbal content of what is being said. It should incorporate all the other things being communicated, for instance the variation in social status of the participants, their emotional attitudes, the various non-verbal signs, such as eye gaze, body signals and so on, the social context of the communication and each individual's perception of the other.

Features of verbal communication

Interpersonal verbal communication is characterized by:

- the way in which the words are spoken
- the accompanying non-verbal information
- facial expression, gestures and posture
- the expectation of participants
- the context in which the transaction occurs.

It incorporates a number of features that assist in the actual communication process. These include:

- reflexes, such as coughing
- non-verbal noises, such as grunts, that need non-verbal accompaniment in order to clarify their meanings
- voice qualities, such as accents, that are significant in categorizing the individual and the social group to which a person belongs
- linguistic aspects, in terms of the choice of words, and paralinguistic aspects, namely timing, speech, rhythm, tone of voice and pitch of voice.

Interpersonal communications thus consist of a complicated fabric of interacting 'cues' or signals of many kinds, in particular the sequence of words coloured by voice, tone, pitch, stress or rhythm, and movement of eyes, hands and body.

Communications almost always requires the use of a 'code' of some kind, such as language, non-verbal cues, emotional states, relationships, understandings.

Language

There are three kinds of meaning attached to words:

- *denotive* the key features that distinguish it
- *connotive* varies according to experience, association and context
- *indexical* provide an indication of the nature of the speaker.

Understanding

This is a critical factor in the communication process. Vital information with respect to hazards, for instance, should be repeated at intervals. The *FIDO* principle is worth noting in this regard. According to this learning by communication is enhanced by:

- *Frequency*
- *Intensity*
- *Duration*
- *Over* again.

Fundamentally, this means that:

- the manager should make sure to tell employees what they need to know – not leave it to them to 'read his mind' or to 'pick up' the necessary facts – and ensure that they are promptly and accurately informed of matters relevant to their work
- communication should be dispensed in small 'doses' as most people can absorb only a limited amount of information at one time (long and involved communications are seldom read or listened to and, if they are, they are rarely digested, so only a few, important points should be communicated at a time)
- the manager should learn to phrase communications in a simple, direct style and consideration must be given to the level of education and experience of employees and others, but even well-educated employees are more likely to perceive the intended message correctly if it is phrased in the most straightforward manner.

On this last point, individuals differ widely in ability. Some managers straightforwardly succeed in making themselves easy to understand, whereas others make the task unbearably difficult. Research workers in language and psychology have studied ths problem in great detail. The results of one study are described in *The Art of Plain Talk* (1946) by Rudolf Flesch. Flesch analysed the elements of language expression that, in general, make for ease of comprehension. These are:

- *number of words in a sentence* generally, the shorter the sentence, the more easily it will be comprehended

- *number of syllables in a word* the shorter the words used – measured by syllables, not necessarily letters – the easier they are to understand

- *number of personal words and sentences* the greater the percentage of personal words and/or personal sentences, as distinguished from impersonal or abstract constructions, the more comprehensible is the language.

6

Barriers to verbal communication

A number of pitfalls can hinder effective verbal communication. For instance:

- the communicator may be unable to think clearly
- there may be problems or difficulties in encoding the message on the part of receivers
- transmission of the message can be interrupted by noise or distractions
- the receiver may exercise selectivity in reception, interpretation and retention of the information in the message
- the receiver may simply not be listening to the message
- an unsuitable environment, such as a workshop, construction site, will act as an impediment to good communication
- a misunderstanding of feedback from the receiver of the message can arise
- there may be no perceptible reaction from the receiver

- rumour can fill the gap in the formal communication system, which is normally associated with the grapevine.

Non-verbal communication

Non-verbal communication is an important feature of the total communication process. It has several functions:

- non-verbal communication can give support to verbal communication in that:
- gestures can add to or emphasize words
- terminal glances help with speech synchronization
- tone of voice and facial expression indicate the mood in which remarks are intended to be taken
- feedback on how others are responding to what is being said is obtained by non-verbal devices, such as facial expressions

- it can replace speech where speech is not possible
- it can perform ritualistic functions in everyday life and can communicate complicated messages in greeting and farewell ceremonies
- it can express feeling we have about others, such as like or dislike
- it can express what condition we are in or feelings we have about others – happiness, anger, anxiety – although we may attempt to control them
- it can be used to convey how we would like other people to see us, for example, by the way in which we present ourselves for public scrutiny.

Aspects of non-verbal communication
These include the following:

- *visual aspects* these aspects include those actually seen during the communication process, such as:
- involuntary (non-controllable) features, such as blushing, pallor, perspiration
- physical appearance, such as gestures, facial expression, gaze
- posture, such as static, movement and change of posture
- orientation

– proximity or closeness between the communicators

● *tactile aspects* the tactile aspects of communication are associated with touch and physical contact, for example:
– aggression, such as hitting and striking

– caressing and stroking

– guiding

– greeting (often formalized)

● *olfactory aspects* these are associated with how people smell

● *auditory aspects* these features of communication are significant. They are concerned with how people hear things and their non-verbal responses.

COMMUNICATIONS FAILURE

Failures in the communication process are commonly associated with:

● *time defects* there may be a time lag between the communication going out and the recipient receiving it

● *spacial segregation* where the communicators may be some distance apart physically

● work groups who, for a variety of reasons, fail to *cordinate* their various activities

● various degrees of *conflict* between the organization's staff, specialists and line management.

Lack of understanding commonly arises as a result of failure to communicate accurately and fully the circumstances surrounding a situation.

COMMUNICATION ON HEALTH AND SAFETY ISSUES

Communication of the right kind has a vital part to play in health and safety as a participative process, but what is the right kind? The following aspects of communication are significant.

- *Safety propaganda* The use of posters, films, exhibitions and other forms of repetition of a specific message are important features of the safety communication process. Safety posters should be used to reinforce current health and safety themes, say, the use of eye protection, correct manual handling techniques, and should be changed on a regular basis. To have the most impact, videos and films should be used as part of scheduled training activities, not shown in isolation.

- *Safety incentive schemes* Various forms of planned motivation directed at rewarding good safety behaviour on the basis of formally agreed objectives and criteria have proved successful. Safety incentive schemes should *not* be based on a reduction in accident rates, however, as this can reduce or restrict accident reporting by employees. All safety incentive schemes must be accompanied by efficient communication of results and information surrounding the scheme.

- *Effective health and safety training* Health and safety communications featured in training exercises should incorporate sincerity, authority, confidence, accuracy and humour. Training of speakers and trainers in the various aspects of presentation can be helpful in making training more effective.

- *Management example* This is perhaps the strongest form of non-verbal communication and has a direct effect on attitudes to health and safety held by operators. Management should openly display and promote, where appropriate:
  - a desire to attain a common goal, such as clearly identified and agreed safety objectives
  - insight into ever-changing situations, particularly in the case of potentially hazardous operations
  - alertness to the needs and motives of others
  - an ability to bear responsibility with regard to health and safety procedures
  - competence in initiating and planning action to bring about health and safety improvements
  - social interaction aimed at promoting the health and safety objectives of the organization

- communication, both upwards and downwards, through a wide variety of communication activities
- clear identification by senior management with the health and safety promotional activities undertaken by the organization
- by setting an example to staff, visitors and others at all times on health and safety-related procedures.

Verbal communication on health and safety issues

The legal duties of employers, including managers and supervisors, to inform, instruct and train staff and others – in other words, to communicate – are clearly identified in current health and safety legislation.

Increasingly, supervisors and line managers are required to prepare and undertake short training sessions on health and safety issues for the purpose of briefing staff. The following matters need careful consideration if such training activities are to be successful and get over the appropriate messages to staff.

- a list of topics to be covered should be developed, followed by the formulation of a specific training programme
- training sessions should be frequent, but should not last longer than 30 minutes
- extensive use should be made of visual aids – videos, films, slides, flip charts and so on
- topics should, as far as possible, be of direct relevance to the group
- participation should be encouraged, with a view to identifying possible misunderstandings or concerns that people may have (this is particularly important when introducing a new safe system of work or operating procedure)
- consideration must be given to eliminating any boredom, loss of interest or adverse response from participants, so sessions should operate on as friendly and informal basis as possible and in a relatively informal atmosphere as many people respond adversely to the formal classroom situations commonly encountered in staff training exercises.

WRITTEN COMMUNICATION

This may be of a formal or informal nature. It may take the form of business letters, memoranda and reports.

- *Business letters* Letters are sent to external organizations either giving or requesting information. As a formal means of communication, they may also be sent internally, conveying information or to elicit information. Business letters must be laid out in a logical and concise manner.

- *Memoranda* Generally, memoranda comprise a few paragraphs and are an informal means of communication between members of an organization. They should be simple and to the point.

- *Reports* A report is defined as 'a written record of activities based on authoritative sources, written by a qualified person and directed towards a predetermined group'. Reports tend to be of an impersonal nature. They state the facts or findings of the author of the

---

The presenter
- Appearance
- Body movement
- Firmness
- Distracting habits
- Timing
- Punctuality
- Gestures and mannerisms
- Eye contact
- Empathy
- Pitch, pace and pauses
- Use of notes and visual aids
- Confidence, conviction and sincerity

SMILE!!!

The presentation
- Preparation and rehearsal
- Content
- Word pictures
- Construction – beginning, middle and end
- Vocabulary

The environment
- Room arrangements – layout, seating, tables
- Temperature, lighting, ventilation

Support material and equipment
- Visuals – must be seen, understood and create interest
- Equipment – must be in working order, with spares available

---

- FIG 6.2 Presentation technique – important points for presenter and presentation

report, for example, following an accident investigation, make recommendations and, in some cases, seek approval for, say, expenditure or for certain actions to be taken or not taken.

Reports provide information, formulate opinions and are directed towards assisting people to make decisions.

Reports should follow a logical sequence. They should be written with clarity and may be accompanied by diagrams, tables and photographs. The principal objective of a report is to enable the reader to reach a conclusion as to future action in certain situations. A report should generally terminate with a recommendation or series of recommendations.

CONCLUSION

Communication is the crucial link in successful integrative management. An essential requirement is that the channels of communication remain open and are used. Effective communication is also an important feature of health and safety practice. Lack of communication is commonly a contributory factor in accidents and other forms of adverse incidents. Barriers to effective communication must be overcome by attention to the points raised in this chapter.

The legal duty of managers to communicate with staff and other persons is clearly identified in HSWA and in most of the regulations made under this Act.

Health and safety training

THE LEGAL DUTY TO PROVIDE HEALTH AND SAFETY TRAINING

Section 2 of the HSWA makes the duty of employers 'to provide such information, instruction, training and supervision as is necessary to ensure, so far as is reasonably practicable, the health and safety at work of his employees'.

This duty is extended in subordinate legislation, such as the Control of Substances Hazardous to Health (COSHH) regulations 1988 and, in particular, the MHSW regulations 1992.

Regulation 11(2) of the MHSW regulations specifies actual situations and circumstances where health and safety training *shall* be provided by the employer as follows. Every employer shall ensure that employees are provided with adequate health and safety training.

- on their being recruited into the employer's undertaking
- on their being exposed to new or increased risks because of:
- their being transferred or given a change of responsibilities within the employer's undertaking
- the introduction of new work equipment into or a change respecting work equipment already in use within the employer's undertaking
- the introduction of new technology into the employer's undertaking
- the introduction of a new system of work into or a change respecting a system of work already in use within the employer's undertaking.

In regulation 11(3), the training referred to in paragraph 2 *shall*:

- be repeated periodically where appropriate
- be adapted to take account of new or changed risks to the health and safety of the employees concerned
- take place during working hours.

It should be appreciated that the duty to provide health and safety training is an absolute one, hence the use of the word 'shall' in the wording of the regulation.

Similar duties exist under the common law duties of the employer (Wilsons & Clyde Coal Co. Ltd vs. English, 1938).

TRAINING

Training has been defined by the Department of Employment as:

The systematic development of attitude, knowledge and skill patterns required by the individual to perform adequately a given task or job. It is often integrated with further education.

Systematic training

The term 'systematic' immediately distinguishes this form of development from the traditional approach, which most often consisted of the trainee 'sitting by Nellie' and acquiring, haphazardly, what they could by listening and observation.

Systematic training, in effect, makes full utilization of skills available in training all grades of personnel. It has a number of benefits. It:

- attracts recruits
- achieves the target of an experienced operator's skill in half or a third of the traditional time
- creates confidence in trainees that they can acquire diverse skills through application and training
- guarantees better safety performance and morale
- results in greater earnings and productivity
- results in ease, basic mental security and commitment at work

- excludes misfits and diminishes unrest
- facilitates the understanding and acceptance of change.

Systematic training involves:

- the presence of a competent and trained instructor and suitable trainees
- defined training objectives
- a content of knowledge broken down into learnable sequential units
- a content of skills analysed into elements
- a clear and orderly programme
- an appropriate place in which to learn
- suitable equipment and visual aids
- sufficient time to attain a desired standard of knowledge and competence.

THE TRAINING PROCESS

The training process takes place in a number of clearly defined stages.

1 *Identification of training needs* A training need is said to exist when the optimum solution to an organization's problem is by means of some form of training. For training to be effective, it must be integrated, to some extent, with the selection and placement policies of the organization. Selection procedures must, for instance, ensure that the trainees are capable of learning what is to be taught.

   In accordance with the MHSWR 1992, training needs should be identified to cover:

- *induction training for* new recruits
- *orientation training* of existing employees on, for instance, promotion, change of job, their exposure to new or increased risks, appointment as competent persons, the introduction of new plant, equipment and technology, and prior to the introduction of safe systems of work

- *refresher training* directed at maintaining competence.

2 *Development of a training plan and programme* Training programmes must be coordinated with the current personnel needs of the organization. The first step in the development of a training programme is that of defining the training objectives. Such objectives or aims may best be designed by job specification, in the case of new training, or by detailed task analysis and job safety analysis, in respect of existing jobs.

3 *Implementation of the training plan and programme* Decisions must be made as to the extent of both active and passive learning systems to be incorporated in the programme. Examples of active learning systems are group discussion, role play, syndicate exercises, programmed learning and field exercises, such as safety inspections and audits. Active learning methods reinforce what has already been taught on a passive basis.

Passive learning systems incorporate lectures and the use of visual material, such as films and videos. With a passive learning system, the basic objective must be that of imparting knowledge. The principal advantage of passive learning systems is that they provide frameworks and can be used where large numbers of trainees are involved. It should be incorporated as an initial introduction to a subject in particular, and should include rules and procedures, providing they are relatively simple.

Active learning systems are the most effective form of training once the basic framework is established and there is plenty of time available in the training programme. It is suitable for a subject where there are no 100 per cent correct answers. There is more chance of bringing about attitude change on the part of trainees and the level of interest of trainees is maintained.

4 *Evaluation of the results* There are two questions that need to be asked at this stage:

- Have the training objectives been met?
- If they have been met, could they have been met more effectively?

Operator training, in most industries, will need an appraisal of the skills necessary to perform a given task satisfactorily, that is, efficiently and safely. It is normal, therefore, to incorporate the results of such appraisal in the basic training objectives.

A further objective of, particularly, health and safety training is to bring about long-term changes in attitude on the part of trainees, which must be linked with job performance. Any decision, therefore, as to whether or not training objectives have been met, cannot be taken immediately the trainee returns to work or after only a short period of time. It may be several months or even years before a valid evaluation can be made after continuous assessment of the trainee.

The answer to the second question can only be achieved through *feedback* from personnel monitoring the performance of trainees, and from the trainees themselves. This feedback can be employed usefully in setting objectives for further training, the revision of the training content and analysis of training needs for all groups within the organization.

Transfer of training

One of the most important problems in the psychology of learning is that of the transfer of training. There are basically two types of transfer:

- *positive transfer of training* this occurs where more rapid learning in one situation is achieved because of previous learning in another situation

- *negative transfer of training* in this case, learning is slower because of previous learning in another situation.

Fundamentally, it is vital that people should be capable of transferring knowledge and skills acquired in one situation to another situation. Any training programme, whether directed to health and safety improvement or for other purposes, should endeavour to bring about this positive transfer of training.

TRAINING METHODS AND TECHNIQUES

Methods and techniques of training are directly related to the way we would like people to learn things. A wide range of techniques and methods is available, depending on the learning system, that is active or passive learning.

- *Guided reading* With this method, the trainee is given standard literature or company material, for example, a safe system of work, to read and comment on in a structured situation. For self-motivated trainees, it can be an effective means of knowledge transfer. Guided reading forms an integral part of most training courses.

- *Lectures* A lecture is defines as 'a straight talk or exposition, possibly using visual or other aids, but without group participation other than through questions at the conclusion'. Group participation is, thus, of an auditory nature.

A lecture can appropriately function to:
- indicate rules, policies, regulations and course resources to trainees
- introduce and provide a general survey of a subject, its scope and its values
- provide a brief on procedures to be adopted in subsequent learning activities
- set the scene for a demonstration, discussion or presentation
- illustrate the application of rules, principles or concepts
- recapitulate, add emphasis or summarize.

It must be appreciated that a lecture, as a passive form of training, is only as good as the lecturer's ability to present and maintain the interest of trainees. The following points should be considered:
- communication is largely one way, with little or no interchange between lecturer and trainee
- a lectures are inappropriate for teaching specific skills
- lecture has limited sense appeal, which is chiefly aural
- lecturing encourages passivity among trainees
- it is not easy to gauge reaction in the lecture situation
- effective lecturing is a highly skilled task and, because trainee interest and attention has to be generated by the lecturer, the latter's vocabulary, enthusiasm, planning, speech techniques, class sensitivity and so on are all critical.

- *Demonstration* In a demonstration, the instructor, by actual performance, shows the trainees what to do and how to do it, and with associated explanations, indicates why, when and where it is to be done. A demonstration is generally combined with some other form of training.

- *Guided practice* This can be defined as 'a method in which the trainee has to perform the operation or procedure, being taught under controlled conditions'. There are four basic categories:
- *independent practice* in which trainees set their own pace and work individually
- *controlled practice* in which trainees work together at a pace set by the instructor
- *team performance* this involves a group of trainees performing together as a team
- *coach and pupil* a method requiring paired trainees who perform alternately as trainee and instructor.

The main applications of guided practice are, in general, those of the demonstration. It is normally used as a follow up instruction in teaching manipulative operations and procedures, the functioning and operation of equipment, team skills and safety procedures.

- *Group discussion* Discussion methods are often classified into three categories: directed discussion, developmental discussion and problem-solving discussion. No sharp boundaries exist between them, and they differ in their objectives.
- *Directed discussion* The objective here is to assist trainees to achieve a better understanding and the ability to apply known facts, principles, concepts, policies and procedures, and to provide trainees with an opportunity to applying this knowledge. The instructor attempts to guide the discussion so that the facts, principles and so on are clearly linked and applied.
- *Developmental discussion* The objective is to pool the knowledge and past experiences of the trainees to develop improved or better-stated principles, concepts, policies and procedures. Topics for developmental discussion are less likely to have clear-cut answers than those of directed discussion. The instructor's task is to elicit contributions from members of the trainee group, based on past experience and bearing on the topic in hand, and the aim is for balanced participation.
- *Problem-solving discussion* This form of group discussion attempts to discover an answer to a question or the solution to a problem.

There is no known best or correct solution and the instructor uses the discussion to find an acceptable solution. The instructor's basic functions are to define the problem as they understand it and to encourage free and full participation in a discussion, the goals of which include identifying the real problem, assembling and analysing data, formulating and testing hypotheses, determining and evaluating possible courses of action, arriving at conclusions and making recommendations to support these conclusions.

- *Syndicate exercises* A larger group can be broken into smaller sub-groups for discussion or problem-solving exercises, such as design of a permit to work system, with the instructor available for consultation and guidance. This technique allows for experience sharing, group decision making and discipline in solving a particular problem. Objectives must be clearly established and the reporting back of the syndicate's findings is important.

- *Group Dynamics (T-Groups)* With this training technique, situations develop or are induced in which trainees' behaviour is examined by other trainees. Group behaviour is also examined by the trainees. This technique is very effective in teaching the trainee about their behaviour and the effects it has on others. Knowledge of the trainee's own behaviour and that of the group as a whole increases, together with the skills needed to work together and communicate. Any Group Dynamics session must be well-controlled by a trained instructor.

- *Programmed instruction learning* This is defined as 'a form of instruction/teaching in which the following factors are present:
  - there is a clear statement of exactly what the trainee is expected to be able to do at the end of the programme
  - the material to be learned, which is itemized and tested, is presented serially in identifiable steps and/or frames
  - trainees follow an actual sequence of frames that is determined for them according to their individual needs
  - feedback of the information of the correctness or otherwise of responses is usually given to the trainee before the next frame is presented'.

Programmed learning systems are commonly used where it is not possible for trainees to attend a particular training course or where it is necessary to teach standard procedures or systems to a large number of trainees over a limited period of time.

- *Role-play/simulations* These techniques increase trainee involvement in the learning process by introducing a realistic element into instruction. They present the trainees with a situation that they have to resolve by acting out the roles of those in the situation.

  Role-plays can be a very efficient and interesting form of training, once basic principles of a subject are understood, and is an important feature of management training.

- *Case studies* These are problematic situations that are presented to a group of trainees, the group having to find the best solution, usually in the form of a written report and/or oral presentation following the study.

- *Individual coaching* This entails a one-to-one relationship between trainer and trainee and can be particularly effective, providing the trainer is fully aware of the objectives and adheres to the written training programme provided. Coaching imparts knowledge, develops skills and forms attitudes during informal, but planned encounters between trainer and trainee.

- *Projects* The form taken by projects can vary immensely, but there must be objectives set for the trainee to meet, together with broad guidelines necessary to encourage initiative. A project stimulates creativity, interest and decision making and information on the trainee's knowledge and personality is fed back to the trainer by the way the trainee undertakes the project work.

- *Assignments* An assignment differs from a project in that the task or investigation is undertaken to close guidelines after a briefing session. An assignment encourages learning transfer to the job situation and is a useful test for a trainee. Realistic assignments should be chosen to avoid frustration and loss of confidence by the trainee.

HEALTH AND SAFETY TRAINING

Health and safety training is an ongoing process that should take place at varying stages of an individual's career within an organization and for the reasons quoted earlier in this chapter.

In all cases, management should assess the training needs of the workforce and implement the various training processes (see Figure 7.1).

Induction training

All staff, irrespective of status within the organization, should receive induction training, some of which will be off the job.

Topics for inclusion in induction training are as follows:

- the organization's statement of health and safety policy and the individual responsibilities of all concerned
- procedures for reporting hazards, accidents, near misses and occupational ill-health
- details of hazards specific to the job, the operating instructions and precautions necessary, together with formally written safe systems of work and emergency procedures
- procedures to follow in the event of fire – means of escape, assembly areas, the use of fire appliances – together with procedures to follow when the fire alarm sounds
- safety-monitoring procedures currently in operation and systems for the measurement of health and safety performance
- current welfare arrangements – sanitation, washing facilities, clothing storage, first aid and arrangements for taking meals
- sources of health and safety information
- the role and function of the health and safety specialist, safety representatives and health and safety committee
- the purpose of the correct use of personal protective equipment where issued.

Orientation training

This form of training can take place for a number of reasons, for example, on promotion to ensure understanding of new duties, on the introduction of new technology, work equipment, substances and/or

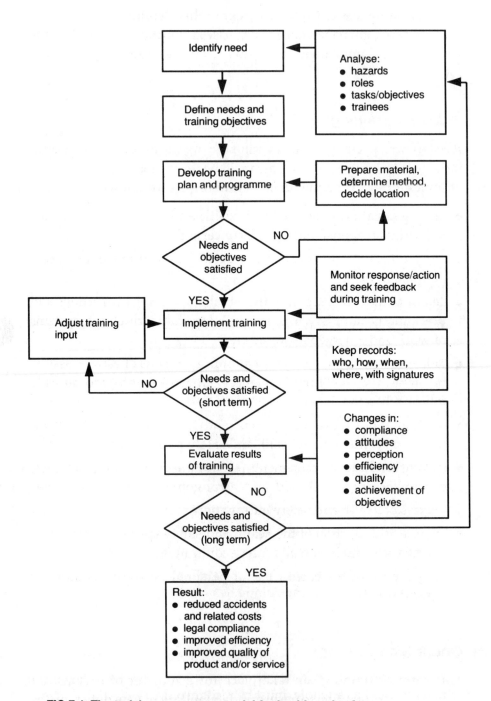

- **FIG 7.1 The training process: a model for health and safety**
(J. Stranks *The Handbook of Health and Safety Practice (3rd ed.)*, Pitman Publishing, 1994.)

systems of work, or prior to working in a new workplace.

Topics for inclusion in orientation training include:

- details of hazards specific to the job and the workplace
- procedures to follow in the event of fire and other emergencies
- safety monitoring systems in operation.

The actual extent of orientation training will vary according to the perceived needs of the individual or group and the hazards to which they may be exposed.

Refresher training

The ACOP to MHSW Regulations makes the following points:

All employee's competence will decline if skills such as those used in emergency procedures are not used regularly. Training, therefore, needs to be repeated periodically to ensure continued competence. Information from personal performance monitoring, health and safety checks, accident investigations and near miss incidents can help to establish a suitable period for retraining.

Special attention should be given to employees who occasionally deputize for others. Their skills are likely to be underdeveloped and they may need more frequent refresher training.

THE BENEFITS TO THE ORGANIZATION

The identification of training needs, particularly in relation to occupational health and safety, is an important task in any organization.

Health and safety training should emphasize certain basic themes, including:

- The importance of distinguishing between:
- accident and injury
- prevention and protection

- the link between safety performance and operational efficiency, and that between accident prevention management and company management as a whole, must be recognized

- training must be regarded as the creation of learning situations, not simply by systematic instruction, but by better design of organizational structures and the adoption of appropriate management styles, so that organizations are viewed as learning systems, which applies in all situations, including that of health and safety at work

- the concept of training must be extended to developing people's full potential so that the organization can satisfy human needs by effective utilization of human resources (it has been said that `accidents downgrade the system', so, in order to ensure effective utilization of human resources, there is clearly a need for regular health and safety training of all staff, from the Board downwards)

- health and safety training should be given at each stage of a person's career, for example at induction stage, prior to the introduction of new processes, substances, work equipment and legal requirements, on promotion, to advise and inform of new responsibilities and on a refresher basis to keep individuals up to date on new legal requirements, procedures, systems, including emergency procedures and new technology

- accidents, occupational ill-health and incidents resulting in property damage and interruption of the business operations represent substantial losses to an organization and a well-trained organization should be able to reduce these losses substantially.

CONCLUSION

Health and safety training is a significant feature of the accident prevention process.

It has been established that the organizations that undertake regular health and safety training have a better safety record, better-informed staff and a reduced potential for disputes. Health and safety training of staff and others, such as employees of contractors, is a legal requirement and should be a feature at all stages of an individual's career within an organization.

Ergonomics

INTRODUCTION

The term *Ergonomics* can be defined in a number of ways:

- the scientific study of the interrelationships between people and their work
- the study of the relationship between man, the equipment with which he works and the physical environment in which this 'man–machine system' operates.

Ergonomics, frequently referred to as 'human factors engineering' or 'the scientific study of work', fundamentally seeks to create working environments in which people receive prime consideration.

The study of ergonomics, which means 'fitting the task to the individual, involves a number of disciplines, including physiology, anatomy, psychology, environmental factors and various areas of engineering. Furthermore, this man–machine interface is significant in the design of work layouts, displays and controls on machinery and equipment, safe systems of work and in the setting of work rates.

The ergonomic approach to health and safety examines, in particular, the physical and mental capacities and limitations of operators, taking into account, at the same time, psychological factors, such as learning, individual skills, perception, attitudes, vigilance, information processing and memory, and physical factors, such as strength, stamina and body dimensions. It is also concerned with the working environment and the potential for environmental stress associated with, for instance, extremes of temperature, inadequate lighting, noise and vibration.

Ergonomics, therefore, is a very broad area of study largely concerned with maximizing human performance and, at the same time, eliminating, as far as possible, the potential for human error.

THE PRINCIPAL FEATURES OF ERGONOMIC CONSIDERATION

- *The human system* All people are different in terms of their physical and mental capacities. This is particularly apparent when considering the physical elements of body dimensions, strength and stamina, coupled with the psychological elements of learning, perception, personality, attitude, motivation and reaction to given situations. Other factors that have a direct effect on performance include the level of knowledge and the degree of training received, their own personal skills and experience of the work.

- *Environmental factors* This area examines the effects on the human system of the working environment with regard to, for instance, the layout of the working area and the amount of work space available and, in particular, the effects on people of environmental stressors, such as extremes of temperature, lighting, ventilation and humidity levels, noise and vibration (see Environmental factors later in this chapter).

- *The man–machine interface* The man–machine interface involves the passing of information from the machine to the operator through the various display elements associated with the machine, such as a pressure gauge. Similarly, commands are given to the machine by means of various controls, such as a foot pedal. The study of display, controls and other design features of machinery, vehicles, automation and communication systems, with a view to reducing operator error and stress on the operator, is a significant feature of ergonomics.

    Factors such as the location, reliability, ease of operation and distinctiveness of controls and the identification, ease of reading, sufficiency, meaning and compatibility of displays are all important in ensuring that machinery is operated correctly.

    So, how do people compare with machines in their ability to perform tasks?

  - *Data processing* In processing data (remembering and interpreting information), people are generally superior to machines in that they do not need extensive programming as does a computer. People are more flexible and can deal readily, in most cases, with unexpected and unforeseen situations such as a fault in a circuit. They can exercise judgement and quickly recall facts and methods of problem solving. However, machines are superior to people in terms of the amount of detailed information that they can store or remember, in the speed and accuracy with which they can provide answers, in classifying and sorting information, in giving reliable results in routine operations, and in working longer at high speed without being subject to fatigue or other factors that distort judgement and decision.

– *Sensory function* People as sensors are restricted in the range of the spectrum of light or sound to which they respond, whereas machines can be designed to sense signals of which people are completely unaware. On the other hand, human sensitivity to many forms of physical energy is exceedingly acute and often better than sensing devices. Furthermore, people's senses operate through a much wider range of intensities, giving good performance for very weak as well as strong stimuli, than do sensing devices.

– *Controlling* People are generally inferior to machines when it comes to controlling things, and the controls they use must be designed to take account of both people's physical and mental limitations. People are relatively weak and slow, limited in the kinds of movements they can make and in the number of controls they can operate simultaneously or in quick succession. The amount of time they can work without fatigue or loss of attention is relatively short. On this basis the physical and mental limitations of people must be considered in the design of tasks.

See also Figure 8.1.

- *Task characteristics* All tasks, no matter how simple they may appear, incorporate a number of characteristics. These may include the frequency of operation of, for instance, a handwheel, the repetitiveness of the task, the actual workload, how critical it is that the task is accurately completed according to a prescribed procedure, the duration of the task and its interaction with other tasks as part of a manufacturing process.

- *Task demands* The specific characteristics of a task will make a number of physical and mental demands on the operator. Certain tasks may require a high degree of physical strength and stamina, such as manual handling activities. Other tasks, such as inspections, may require a high degree of attention and vigilance in ensuring that products not conforming to specification are removed from the process at a particular stage. Most tasks require some form of memory utilization.

Advantages	Disadvantages
People	*Machines*
● Adaptable and flexible	● Relatively inflexible
● Can detect minute stimuli and assess small changes	● Can detect if programmed, but not assess
● Can interpolate and use judgement	● Can do neither
● Can synthesize and learn from experience	● Can do neither
Machines	*People*
● Can operate in hostile environments	● Lower capability
● Fast response to emergency signals	● Slow response
● Can apply large forces smoothly	● Can apply large forces coarsely
● Information storage: large short-term memory	● Easily distracted: limited short-term memory
● Perform routine repetitive task reliably	● Not reliable for this
● Compute fast and accurately	● Compute slowly and inaccurately
● Can operate for long periods without maintenance	● Suffer relatively soon from fatigue and monotony

● **FIG 8.1 A comparison between people and machines**

- *Instructions and procedures* The quality of both verbal and written instructions to operators and formal work procedures have a direct relationship with the potential for human error. As such, instructions and written procedures should be clear, unambiguous, sufficient in detail, readily understandable, easy to use, accurate and produced in an acceptable format. They should be subject to regular revision, particularly where operators are experiencing difficulties with their interpretation and use.

- *Potential stressors* Many tasks, due to a failure to consider the physical and mental needs of the operator – level of intelligence, attitude and motivation – can be stressful. The following questions should be asked during the design of tasks and at the task analysis stage.

- *Isolation* Does the task isolate the operator, both visually and audibly, from fellow operators? Isolation from the working group can be a significant cause of stress.
- *Time pressures* Does the task put operators under pressure due to the need to complete it within a certain time scale? Such situations create stress and increase the potential for human error.
- *The workload* Does the task impose a higher level of mental and physical workload on people than normal? Current sickness absence levels may be a direct indicator of an overload situation.
- *Monotony* Is the task of a highly repetitive nature? Monotony and boredom are a standard feature of many production processes. While many operators are quite happy to undertake repetitive tasks, others will find these tasks stressful, resulting in fatigue. It is possible to rotate operators on different tasks to reduce the monotony and boredom?
- *Conflict* Do the various tasks create conflict among operators? Conflict can arise as a result of varying workloads and levels of complexity for different tasks. Similarly, rates of pay may vary according to the significance of individual tasks, resulting in conflict over the allocation of the more highly paid tasks.
- *Pain and discomfort* Do some tasks result in physical pain or discomfort, such as manual handling operations or working in low temperatures? These factors should be considered as part of job safety analysis and in the design of safe systems of work.
- *Distractions* Is there a risk of distraction during certain critical tasks? Distraction is one of the principal causes of people making mistakes, particularly in various forms of assembly work, the results of which could be significant. Distraction will also increase stress on operators, due to the need to regain concentration while carrying out these highly critical tasks. Critical tasks should be planned with consideration being given to the potential for distraction. In certain cases, design of the working area should be such as to prevent people from being distracted.
- *Space allocation* Is there sufficient space available to undertake the work safety? Congested, badly planned work area layout can alone create stress and increase the potential for mistakes and accidents.

- *Shift work* Where a shift work system is in operation, does this system take account of the physical and mental limitations of operators? Working long hours on rotating shift systems can have a direct effect on individual behaviour. While many individuals can cope satisfactorily with shift work, others find it stressful for different reasons.
- *Incentives* Where an incentive scheme is in operation, are the incentives offered seen as fair to all concerned? Incentive schemes frequently create conflict in that certain groups of workers are seen to benefit more than others.

- *The socio-technical factors* The socio-technical factors cover a wide range of considerations. They include the social relationships between operators and how they work together as a group, group working practices, manning levels, working hours and the provision of meal and rest breaks, the formal and informal communication systems and the rewards and benefits available. Organizational features are also incorporated under this heading, such as the actual structure of the organization and individual work groups, the allocation of responsibilities, the identification of authority for certain actions and the interfaces between different groups.

The total working systems

Taking into account the factors mentioned above, it is quite common for the approach to ergonomics to be summarized under the concept of the *Total Working System*, which is broken down into the four major elements shown in Figure 8.2. There is no doubt that ergonomic considerations make a great contribution to safety and accident prevention.

DESIGN ERGONOMICS

Design ergonomics is a branch of ergonomic study concerned with the design and specification of the various features of the man–machine interface. Important features in the design of this man–machine interface include the following.

- *Layout* The layout of working areas and operating positions should allow for free movement, safe access and egress, and unhindered visual and oral communication. Congested, badly planned layouts result in operator fatigue and increase the potential for accidents.

- *Vision* The operator should be able to set controls and read dials and displays with ease. This reduces fatigue and accidents arising from faulty or incorrect perception.

Human characteristics:
- body dimensions
- strength
- physical and mental limitations
- stamina
- learning
- perception
- reaction

Man–machine interface:
- displays
- controls
- communications
- automation

Environmental factors:
- temperature
- humidity
- lighting
- ventilation
- noise
- vibration

Total working system:
- fatigue
- work rate
- posture
- productivity
- accidents
- safety aspects
- occupational ill-health

- **FIG 8.2 The Total Working System**

- *Posture* The more abnormal the working posture, the greater the potential for fatigue and long-term injury. Work processes and systems should be designed to permit a comfortable posture that reduces excessive job movements. This must be considered in the siting of controls and in the organization of working systems, such as assembly or inspection work.

- *Comfort* The comfort of the operator, whether driving a vehicle or operating machinery, is essential for their physical and mental well-being. Environmental factors directly affect comfort and should be given priority.

Principles of interface design

- *Separation* Physical controls should be separated from visual displays. The safest routine is achieved where there is no relationship between them.

- *Comfort* If separation cannot be achieved, control and display elements should be mixed to produce a system that can be operated with ease.

- *Order of use* Controls and displays should be set in the order in which they are used, for example, left to right for start-up and the reverse direction for close-down.

- *Priority* Where there is no competition for space, the controls most frequently used should be sited in key positions. Controls such as emergency stop buttons should be positioned where they can be most easily seen and reached.

- *Function* With large consoles, controls can be divided according to functions. Such a division of controls is found in power stations. This layout relies heavily on the operator and their speed of reaction. A well-trained operator, however, benefits from such functional division and the potential for error is greatly reduced.

- *Fatigue* The convenient siting of controls is paramount. In designing a layout, the hand movements and body positions of the operator should be observed and studied with a view to reducing or minimizing excessive movements.

See also Figure 8.3.

ANTHROPOMETRY

This is the study and measurement of body dimensions, the orderly treatment of resulting data and the application of this data in the design of workplace layouts and equipment.

This need to match the physical dimensions of people with the equipment they use was aptly demonstrated by the Cranfield Institute of Technology, who created 'Cranfield Man'. Studying a horizontal lathe, researchers examined the positions of controls and compared the locations of these controls with the physical dimensions of the average operator. Figure 8.4 and Table 8.1 show the

Display type	Functional characteristics			
Qualitative				
Auditory	For attracting immediate attention/ Annunciators are warning devices.			
Visual	For representing three or more status conditions by the use of colour, shape, size or location of coding.			
Representational	Provides operator with a model of the system. Good for showing spatial relationships between variables. A static example is a map, a dynamic one, process control.			
Quantitative				
	Ease of reading	*Precision*	*Directing rate of change*	*Setting to a reading*
Analogue (dials) and meters)				
– moving pointer	Acceptable	Acceptable	Very good	Very good
– moving scale	Acceptable	Acceptable	Acceptable	Acceptable
Digital	Very good	Very good	Poor	Acceptable

- **FIG 8.3 Functional characteristics of the main types of display**
  (E. Grandjean, *Fitting the Task to the Man*: An ergonomic approach, Taylor & Francis, 1980)

8

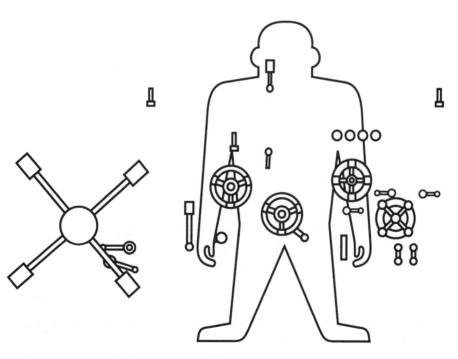

- **FIG 8.4 'Cranfield Man' – 1.35m tall with a 2.44m arm span**

TABLE 8.1 The physical dimensions of the average operator compared with those of 'Cranfield Man'

Average Operator	Dimensions	Operator who would would suit these controls ('Cranfield Man')
1.75m	Height	1.35m
0.48m	Shoulder width	0.61m
1.83m	Arm span	2.44m
1.07m	Elbow height	0.76m

wide differences there were between the two sets of dimensions, or which would clearly result in fatigue and an increased potential for error on the part of the operator.

Anthropometry has also been successfully employed in the design and location of physical guarding to machinery, particularly regarding specifying the distances at which physical barriers should be erected from identified danger points and permissible openings in fixed guards.

ENVIRONMENTAL FACTORS

The design and control of the working environment is an important feature of ergonomics. Environmental working conditions have a direct effect on human behaviour at work, the degree of risk of occupational disease and/or injury, and on morale, management/worker relations, labour turnover and profitability.

Under section 2(2)(e) of HSWA, an employer has a duty to provide and maintain a working environment for all employees that is, so far as is reasonably practicable, safe and without risks to health and adequate as regards facilities and arrangements for their welfare at work. More specific duties employers have are detailed in the Workplace (Health, Safety and Welfare) Regulations 1992 (the 'Workplace' Regulations).

Fundamentally, two aspects must be considered, namely:

- the organization of the working environment
- the prevention or control of environmental stressors.

The organization of the working environment

- *The location of workplaces* Consideration should be given, in the case of large undertakings, to the locality, the density of surrounding buildings, availability of vehicle parking areas, access for employees and transport, including employees' own vehicles, and for fire, ambulance and police vehicles.

  The need for a well-organized traffic control system that does not expose pedestrians to risk of injury must also be considered.

- *Layout* The term 'layout' refers to the space available for those employees working within a particular room or area, and the situation of plant, equipment, machinery, furniture and stored goods in relation to the employee and the tasks performed.

  An efficient layout should make a material contribution to preventing or reducing overcrowding, minimizing the physical and mental effort required to undertake the operations and expediting the work process in an orderly and sequential flow, ensuring, at the same time, maximum safety and hygiene standards throughout.

  A well-planned layout should eliminate the problem of overcrowding. Part 1 of Schedule 1 of the workplace regulations deals with the question of space availability thus:
  - no room in the workplace shall be so overcrowded as to cause risk to the health or safety of persons at work in it
  - without prejudice to the generality of paragraph 1, the number of persons employed at a time in any workroom shall not be such that the amount of cubic space allowed for each is less than 11 cubic metres
  - in calculating, for the purposes of this part, the amount of cubic space in any room, no space more than 4.2 metres from the floor shall be taken into account, and where a room contains a gallery, the gallery shall be treated for the purposes of this schedule as if it were partitioned off from the remainder of the room and formed a separate room.
- *Structural safety* Attention must be paid to the structural safety

aspects of:
- floors, corridors and passageways
- stairs, ladders and catwalks
- underground rooms
- external areas and approach roads.

- *Cleanliness and waste management* Regulation 9 of the Workplace Regulations requires that every workplace and the furniture, furnishings and fittings therein shall be kept sufficiently clean. The surfaces of the floor, wall and ceiling of all workplaces inside buildings shall be capable of being kept sufficiently clean and, so far as is reasonably practicable, waste materials shall not be allowed to accumulate in a workplace except in suitable receptacles.

  In line with the general duty of employers under the MHSW to manage health and safety activities, management should manage cleaning by operating formal cleaning schedules or programmes that take account of waste storage and removal requirements.

- *Colour* Colour is an important factor in the maintenance of a safe working environment. It influences the extent to which the creation of a congenial environment is achieved, the efficiency of lighting and general safety performance. Safety signs should be provided in accordance with the shapes and colours required under the Safety Signs Regulations 1980.

The prevention or control of environmental stressors

Environmental stress associated with, for example, poor lighting levels, extremes of temperature and humidity, inadequate ventilation and noise and vibration, is a contributory factor to both accidents and occupational ill-health.

Temperature

The workplace regulations require that the temperature during working hours in a workplace be 'reasonable'. This implies the need to consider the type of work being carried out – active or sedentary work – and the season of the year in determining what is a reasonable temperature in terms of comfort. Recommended working temperatures are shown in Table 8.2.

Ventilation

Two aspects must be considered here, namely:

- the provision of adequate 'comfort' ventilation in terms of sufficient quantity of fresh or purified air
- means for the removal of airborne contaminants, such as dusts, gases, vapours or fumes, by the operation and maintenance of local exhaust ventilation systems.

TABLE 8.2 Recommended working temperatures

Type of work	Temperature
● Sedentary/office work:	
– comfort range	19.4°C to 22.8°C
● Light work:	
– optimum temperature	18.3°C
– comfort range	15.5°C to 20°C
● Heavy work:	
– comfort range	12.8°C to 15.6°C

TABLE 8.3 Recommended comfort ventilation rates

Location	Summer	Winter
Offices	6	4
Corridors	4	2
Amenity areas	6	4
Storage areas	2	2
Production areas with heat-producing plant	20	20
Production areas (assembly, finishing work)	6	4
Workshops	6	4

Ventilation systems, whether natural or artificial, should be designed assuming a maximum air temperature of 32.2°C and a minimum air temperature of 0°C and should operate, from a comfort viewpoint, to give the specific number of air changes per hour shown in Table 8.3.

The following points need to be considered in the design and specification of comfort ventilation systems:

- they need to produce an atmosphere that is cool rather than hot, dry rather than damp, moving rather than still, with relative humidity between 40 and 70 per cent

- incoming air should be drawn from a clean source or should be filtered

- every room, passage and staircase should be separately ventilated (door openings should not be included in ventilation calculations)

- heat should be removed as close to the source of emission as possible

- inputs should be sited to give a flow of air from operating positions towards heat sources and thence to extracts

- fresh air intakes on roofs should stand at least 700mm clear of the roof surface to avoid picking up heated air from exhausts

- the total extract volume should be 80 per cent of the input volume to allow for a positive plenum, that is keeping the building under pressure.

Lighting

Every workplace must have 'suitable and sufficient' lighting – 'suitable' in terms of the qualitive factors, such as distribution, brightness, diffusion and freedom from glare, and 'sufficient' as far as the actual amount of light, measured in lux, is concerned for both general areas and specific tasks.

The HSE's guidance note HS(G)38 'Lighting at work', provides information on the quantitative aspects of lighting, distinguishing between 'average illuminance' and 'minimum measured illuminance' according to the general activity undertaken and the type of location and work carried out. This information is detailed in Table 8.4. The note also provides guidance on the ratio of illumination between working areas and adjacent illuminance areas (see Table 8.5).

In any assessment of lighting provision, it is not sufficient merely to measure illumination with a standard photometer (light meter), it is also necessary to consider the qualitative features of the lighting installation. On this basis, the following features need to be considered.

- *Glare* This is the light effect that causes discomfort or impaired vision,

TABLE 8.4 Average illuminances and minimum measured illuminances for different types of work

General activity	Typical locations/ types of work	Average illuminances in lux	Minimum measured illuminces in lux
Movement of people, machines and vehicles (1)	Lorry parks, corridors, circulation routes	20	5
Movement of people, machines and vehicles in hazardous areas; rough work not requiring any perception of detail (1)	Construction site clearances, excavation and soil work, docks, loading bays, bottling and canning plants	50	20
Work requiring limited perception of detail (2)	Kitchens, factories assembling large components, potteries	100	50
Working requiring perception of detail (2)	Offices, sheet metal work, bookbinding	200	100
Work requiring perception of fine detail (2)	Drawing offices, factories assembling electronic components, textile production	500	200

Notes: 1 Only safety has been considered, because no perception of detail is needed and visual fatigue is unlikely. However, where it is necessary to see detail, to recognize a hazard or where error in performing tasks could put someone else at risk, for safety purposes as well as to avoid visual fatigue, the figure should be increased to that for work requiring the perception of detail.

2 The purpose is to avoid visual fatigue; the illumances will be adequate for safety purposes.

and is experienced when parts of the visual field are excessively bright compared with the general surroundings. This usually occurs when the light source is directly in line with the visual task or when light is reflected off a given surface or object. Glare is experienced in three different forms:

- *disability glare* is the visually disabling effect caused by bright bare lamps directly in the line of sight and the resulting impaired vision (dazzle) may be hazardous if experienced when working in high-risk processes, say, at heights or when driving
- *discomfort glare* is caused mainly by too much contrast of brightness

Situations to which recommendation applies	Typical location	Maxmim ratio of illuminances	
		Working area	Adjacent area
Where each task is individually lit and the area around the task is lit to a lower illuminance	Local lighting in an office	5 :	1
Where two working areas are adjacent, but one is lit to a lower illuminance than the other	Localized lighting in a works store	5 :	1
Where two areas are lit to different illuminances by a barrier, but there is frequent movement between them	A storage area inside a factory and a loading bay outside	10 :	1

TABLE 8.5 Maximum ratios of illuminance for adjacent areas

between an object and its background, and is associated with poor lighting design. It causes visual discomfort without necessarily impairing the ability to see detail, but over a period, can cause eye strain, headaches and fatigue. Discomfort glare can be reduced by the careful design of shades that screen the lamp by keeping luminaires as high as practicable and by maintaining luminaires parallel to the main direction of lighting.

– *reflected glare* is the reflection of bright light sources on shiny or wet work surfaces, such as glass or plated metal, which can almost entirely conceal the detail in or behind the object that is glinting (care is necessary in the use of light sources of low brightness and in the arrangement of the geometry of the installation to avoid glint at the particular viewing position).

● *Distribution* The distribution of light, or the way in which light is spread, is important in lighting design. Poor distribution may result in the formation of shadowed areas, which can create dangerous situations, particularly at night. For good general lighting, regularly spaced luminaires are used to give evenly distributed illumination.

This evenness depends on the height of the luminaires above the working position and the spacings of fittings.

- *Colour rendition* This refers to the appearance of an object under a given light source, compared to its colour under a reference illuminant, such as natural light. Colour rendition enables the colour appearance to be correctly perceived. The colour-rendering properties of light fitments should not clash with those of natural light and should be equally effective at night when no daylight contributes to the illumination of the workplace.

- *Brightness* Brightness, or luminosity, is, essentially, a subjective sensation and cannot be measured. It is possible to consider, however, a *brightness ratio*, which is the ratio of apparent luminosity between a task object and its surroundings. To achieve the recommended *brightness ratio*, the reflectivity of all surfaces in the workplace should be carefully maintained and consideration given to reflectance values in the design of interiors. Given a task illuminance factor of 1, the effective reflectivity values should be:
  - ceilings, 0.6
  - walls, 0.3 to 0.8
  - floors, 0.2 to 0.3.

- *Diffusion* This is the projection of light in many directions with no one direction being predominant. Diffused lighting can soften the output from a particular source and so limit the amount of glare that may be encountered from bare fittings.

- *Stroboscopic effects* All lamps that operate from an alternating current supply produce oscillations in light output. When the magnitude of the oscillations is great and their frequency is a multiple or submultiple of the frequency of movement of machinery, the machinery will appear to be stationary or moving in a different manner. This is called a *stroboscopic effect*. It is not common with modern lighting systems, but, where it does occur, it can be dangerous, so appropriate action should be taken to avoid it. Possible remedial measures include:
- supplying adjacent rows of light fittings from different phases of the electricity supply
- providing a high-frequency supply
- washing out the effect with local lighting that has much less variation in light output, such as tungsten lamps

– using high-frequency control fittings if applicable.

Noise and vibration

Exposure to noise above 90 decibels can cause occupational deafness. Excessive noise can distract attention and affect concentration, mask audible warning signals or interfere with the work process, thereby becoming a contributory factor in accidents.

Unnecessary noise should be prevented or controlled. The sort of control of noise needed depends on the actual source of the noise, such as machinery, and the pathway taken by the noise to the recipient. Methods of noise control are summarized in Table 8.6.

Table 8.6 Methods of noise control

Sources and pathways	Control measures
Vibration produced by machinery operation	Reduction at source, say, substituting nylon components for metal, tapered tools on presses
Structure-borne noise (vibration)	Vibration isolation, such as resilient mounts and connections, anti-vibration mounts
Radiation of structural vibration	Vibration damping to prevent resonance
Turbulence created by air or gas flow	Reduction at source or use of silencers
Noise taking an airborne pathway	Noise absorption – no reflection; use of porous lightweight barriers

MANUAL HANDLING OPERATIONS

More than a quarter of the accidents reported each year to the health and safety enforcing authorities are associated with manual handling – the transporting or supporting of loads by hand or by bodily force. The vast majority of reported manual handling accidents result in injuries that take over three days to recover from, most commonly a sprain or strain, often of the back (HSE, 1992).

Approximately 13 per cent of all certified sickness absence in the UK is due to back pain. The principal causes of back pain stem from:

- heavy manual work – lifting and handling, forceful exertion, bending, twisting and so on
- working in a stooped position
- prolonged sitting in a fixed position
- vibration
- psychological stress

as P. Pheasant and D. Stubbs report in *Lifting and Handling: An ergonomic approach* (The National Back Pain Association, 1992).

Manual handling injuries and conditions

Manual handling injuries and conditions can be of an external or internal nature. External injuries include cuts, bruises, crush injuries and lacerations to fingers, hands, forearms, ankles and feet. Generally, these injuries are not as serious as the internal forms of injury, which include muscle and ligamental tears, hernias, prolapsed intervertebral discs and damage to knee, ankle, shoulder and elbow joints. Rheumatism, osteoarthritis and lumbago are conditions that commonly result from manual handling injuries.

Legal requirements for manual handling

The Manual Handling Operations Regulations 1992 establish a clear hierarchy of measures that must be followed by employers. Under these regulations, employers are required to:

- avoid hazardous manual handling operations so far as is reasonably practicable, which may be done by redesigning the task to avoid moving the load or by automating or mechanizing the process
- make a suitable and sufficient assessment of any hazardous manual handling operations that cannot be avoided
- reduce the risk if injury from these operations so far as is reasonably practicable; particular consideration should be given to the provision of mechanical assistance, but where this is not reasonably practicable, then other improvements to the task, the load and the working environment should be explored.

The word *injury* in this context does not include injury caused by any toxic or corrosive substances that:

- has leaked or spilled from a load
- is present on the surface of a load, but has not leaked or spilled from it
- is a constituent part of a load

and the word *injured* shall be construed accordingly. The word *load* is taken to mean any person or animal, while *manual handling operations* means any transporting or supporting of a load (including the lifting, putting down, pushing, pulling, carrying or moving thereof) by hand or bodily force.

Where it is not reasonably practicable to avoid hazardous manual handling operations by employees, an employer must make a suitable and sufficient assessment of all such manual handling operations to be undertaken by them, having regard to the factors that are specified in column 1 of schedule 1 to the regulations and consider the questions that are specified opposite in column 2 of this schedule (regulation 4(1)(b)(i)).

Schedule 1 to the regulations deals with:

Factors to which the employer must have regard and questions he must consider when making an assessment of manual handling operations

Regulation 4(1)(b)(i)

Column 1 *Factors*	Column 2 *Questions*

1 The tasks	Do they involve:
	• holding or manipulating loads at a distance from the trunk?
	• unsatisfactory bodily movement or posture, especially:
	  – twisting the trunk?
	  – stooping?

Column 1	Column 2
-	– reaching upwards?
	• excessive movement of loads, especially:
	– excessive lifting or lowering distances?
	– excessive carrying distances?
	• excessive pushing or pulling of loads?
	• risk of sudden movement of loads?
	• frequent or prolonged physical effort?
	• insufficient rest or recovery periods?
	• a rate of work imposed by a process?
2 The loads	Are they:
	• heavy?
	• bulky or unwieldy?
	• difficult to grasp?
	• unstable, or with contents likely to shift?
	• sharp, hot or otherwise potentially damaging?
3 The working environment	• space constraints preventing good posture?
	• uneven, slippery or unstable floors?
	• variations in level of floors or work surfaces?
	• extremes of temperature or humidity?
	• conditions causing ventilation problems or gusts of wind?
	• poor lighting conditions?
4 Individual capability	Does the job:
	• require unusual strength, height, etc?
	• create a hazard to those who might be pregnant or have a health problem?
	• require special information or training for its safe performance?
5 Other factors	Is movement or posture hindered by personal protective equipment or by clothing?

DISPLAY SCREEN EQUIPMENT (VDUs)

The Health and Safety (Display Screen Equipment) Regulations 1992 (the 'DSE Regulations') define 'display screen equipment' as meaning 'any alphanumeric or graphic display screen, regardless of the display process involved'.

The HSE's publication, *Visual display units* (1983), defined the term visual display unit as 'a specific type of device using a light beam or cathode rays to generate an image which may be displayed wholly or partially in letters or figures, or simple graphics, or even as a picture, on the surface of its associated and normally self-contained screen'.

Occupational health aspects

In the last decade, considerable attention has been given to the occupational health risks associated with display screen work. A description of these risks, alleged and otherwise, is given below.

- *Visual fatigue (eye strain)* Eye strain or visual fatigue is a common feature of many tasks. Common symptoms are irritation of the eyes, which is exacerbated by rubbing, redness and soreness, together with temporary blurring and visual confusion. Spots, shapes in front of the eyes and chromatic effects surrounding viewed objects also occur. Some individuals experience photophobia, resulting in them wearing dark glasses. Headaches are the most common symptom, the nature and location of the pain varying from person to person.

  Generally, visual fatigue has a varied and complicated pattern of symptoms. Moreover, it cannot be separated from general fatigue and is as much related to psychological demands as to malfunctioning of the visual system. For instance, anxiety is often a component of visual fatigue. Visual fatigue is a temporary and reversible phenomenon and it is current medical opinion that it is impossible to damage the eyes by using display screen equipment. The factors of age, visual acuity and performance are significant in assessing a person's potential for visual fatigue.

  Visual fatigue, the principal form of operational stress, can be associated with:

  - poor legibility due to factors in the VDU such as 'flicker', 'shimmer' and 'jitter', which are directly related to the refresh rate of the display screen
  - poor definition of characters against the background field
  - glare
  - unsuitable background lighting

– visual defects.

All but the last can be corrected by good VDU and workstation design. Very few people, however, have perfect vision, the ability to see varying with age and the presence or absence of such visual deficiencies as myopia (short-sightedness) and hypermetropia (long-sightedness).

● *Facial dermatitis* The potential for dermatitis is associated with age, sex, race and certain characteristics specific to the individual. In the same way, that only some people will suffer facial dermatitis through exposure to the sun, a very small proportion of the population will experience this condition when working with display screen equipment. Dermatitis is very much a matter for medical examination and treatment. In some cases, it may be a manifestation of stress.

● *Postural fatigue* No two people adopt the same posture when sitting or standing. Much will depend on the nature of the task and the design of the workplace. Postural fatigue can result in back, shoulder and head aches, loss of sensation in various parts of the body and temporary loss of strength in the hands and arms. Correct design of the workstation is, therefore, of utmost significance, together with the provision and maintenance of a suitable working environment.

● *Work-related upper limb disorders* Work-related upper limb disorders caused by repetitive strain injuries (RSI) were first defined in the medical literature by Bernardo Ramazzini, the Italian father of occupational medicine, in the early eighteenth century. The International Labour Organization recognized RSI as an occupational disease in 1960, as 'a condition caused by forceful, frequent, twisting and repetitive movements'. Many people, including assembly workers, supermarket checkout assistants and keyboard operators, are affected by RSI at some point in their lives.

In 1992, a court awarded two British Telecom computer operators damages of £6000 each in respect of musculoskeletal injuries sustained while using a keyboard. The judgment, in the Mayor's and City of London Court, held that both women had received repetitive strain injuries arising from poor posture and intensive, prolonged and repetitive keyboard work.

RSI covers some well-known conditions, such as tennis elbow (flexor tenosynovitis) and carpal tunnel syndrome. It is usually caused or aggravated by work, and is associated with repetitive and over-forceful movement, excessive workloads, inadequate rest periods and sustained or constrained postures, resulting in pain and soreness due to the inflammatory conditions of muscles and the synovial lining of the tendon sheath. Present approaches to treatment are largely effective, provided the condition is treated in its early stages. Tenosynovitis has been a prescribed industrial disease since 1975 and the HSE has changed the name of the condition to 'work-related upper limb disorder on the grounds that the disorder does not always result from repetition or strain, and is not always a visible injury.

Clinical signs and symptoms include local aching pain, tenderness, swelling and crepitus (a grating sensation in the joints), aggravated by pressure or movement. Tenosynovitis affecting the hand or forearm is the most common prescribed industrial disease, the most common being dermatitis. True tenosynovitis, where inflammation of the synovial linking of the tendon sheath is evident, is rare and potentially serious. The more common and benign form is peritendinitis crepitans, which is associated with inflammation of the muscle-tendon joint that often extends well into the muscle.

The various forms of RSI, then, can be classified thus:
– *epicondylitis* inflammation of the area where a muscle joins a bone
– *peritendinitis* inflammation of the area where a tendon joins a muscle
– *carpal tunnel syndrome* a painful condition of the area where nerves and tendons pass through the carpal bone in the hand
– *tenosynovitis* inflammation of the synovial linking of the tendon sheath
– *tendinitis* inflammation of the tendons, particularly in the fingers
– *Dupuytren's contracture* a condition affecting the palm of the hand, where it is impossible to straighten the hand and fingers
– *Writer's cramp* a condition resulting in cramps in the hand, forearm and fingers.

RSI-related injuries can be prevented by:
– improved design of working areas and workstations, such as positions of keyboards and display screens, heights of desks and chairs
– adjustments of workloads and rest periods
– provision of special aids, such as hand/wrist supports
– health surveillance aimed at detecting early stages of the disorder
– better training and supervision.
If untreated, RSI can be seriously disabling.

- *Pregnancy risks, cataracts and epilepsy* There is no conclusive scientific evidence linking display screen work with adverse consequences during pregnancy, such as miscarriages and birth deformities, and incidence of cataracts of the eye or of epilepsy.

Health surveillance

There is clearly a case for vision screening of all display screen operators on a regular basis, both as part of a pre-employment health screen and at specific periods during employment.

Wearers of spectacles designed for a narrow range of reading distances and those with bifocal and multifocal lenses, may find difficulties with tasks involving varying distances. They may find, for instance, that they need to adopt uncomfortable postures in order to view documents or displayed text satisfactorily. These people may need modifications to their prescription lenses to undertake display screen work and should consult an optician prior to undertaking such work or whenever discomfort or visual fatigue is experienced afterwards.

The use of medication, such as mild tranquillizers and other psychoactive drugs, is now quite common. Occasionally, side-effects of these drugs may occur that mimic some of the symptoms of visual fatigue, such as slowing of the eye movements. Display screen users, who may have been prescribed such medication, should be made aware of these side-effects.

The design of workstations

The DSE regulations define a workstation as an assembly comprised of:

- display screen equipment (whether or not it is provided with software determining the interface between the equipment and its operator or user, a keyboard or any other input device)
- any optional accessories to the display screen equipment
- any disk drive, telephone, modem, printer, document holder, work chair, work desk, work surface or other item peripheral to the display screen equipment
- the immediate work environment around the display screen equipment.

A *user* is an employee who habitually uses display screen equipment as a significant part of their normal work. Typical users are data input operators, secretaries, typists, graphic designers, word processing pool workers and journalists.

Legal requirements relating to workstation design are specified in the schedule to the regulations. Further guidance is detailed in Annex A, 'Guidance on workstation minimum requirements'. The principal features of this practical guidance are given below (see also Figure 8.5).

Equipment

- *Display screen* The choice of display screen should be considered in relation to other elements of the work system, such as the type and amount of information required for the task and environmental factors. A satisfactory display can be achieved by custom design for a specific task or environment or by appropriate adjustments to adapt the display to suit changing requirements or environmental conditions.

- *Display stability* Individual perceptions of screen flicker vary and a screen that is flicker free to 90 per cent of users should be regarded as satisfying the minimum requirement (it is not technically feasible to eliminate flicker for all users). A change to a different display can resolve individual problems with flicker. Persistent display instabilities – flicker, jump, jitter or swim – may indicate basic design problems and assistance should be sought from suppliers.

- *Brightness and contrast* Negative or positive image polarity (light characters on a dark background, dark characters on a light background respectively) is acceptable, and each has different advantages. With negative polarity, flicker is less perceptible, legibility is better for those with low acuity vision and characters may be perceived as larger than they are. With positive polarity, reflections are less perceptible, edges appear sharper and lumination balance is easier to achieve.

  It is important for the brightness and contrast of the display to be appropriate for ambient lighting conditions. Trade-offs between character brightness and sharpness may be needed to achieve an acceptable balance. In many kinds of equipment, this is achieved by providing a control or controls that allow the user to make adjustments.

- *Screen adjustability* Adjustment mechanisms allow the screen to be tilted or swivelled to avoid glare and reflections and enable the worker to maintain a natural and relaxed posture. They may be built into the screen, form part of the workstation furniture or be provided by separate screen support devices. They should be simple and easy to operate. Screen height adjustment devices, although not essential, may be a useful means of adjusting the screen to the correct height for the worker.

– *Glare and reflections* Screens are generally manufactured without highly reflective surface finishes, but, in adverse lighting conditions, reflection and glare may be a problem.

• *Keyboard* Keyboard design should allow workers to locate and activate keys quickly, accurately and without discomfort. The choice of keyboard will be dictated by the nature of the task and determined in relation to other elements of the work system. Hand support may be incorporated into the keyboard for support while keying or at rest, depending on what the worker finds comfortable, may instead be provided in the form of a space between the keyboard and the front edge of the desk, or may be given by a separate hand/wrist support attached to the work surface.

• *Desk or other work surface* Work surface dimensions may need to be larger than for conventional non-screen office work to take account of:

– the range of tasks performed (screen viewing, keyboard input, use of other input devices, writing on paper and so on)
– position and use of hands for each task
– use and storage of working materials and equipment (documents, telephones and so on)

Document holders are useful for work with hard copy, particularly for workers who have difficulty in refocussing. They should position working documents at a height, visual plane and, where appropriate, viewing distance similar to those of the screen, be of low reflectance, be stable, and not reduce the readability of source documents.

• *Work chair* The primary requirement here is that the work chair should allow the user to achieve a comfortable position. Seat height adjustments should accommodate the needs of users for the tasks performed. The schedule to the regulations requires the seat to be adjustable in height (that is, relative to the ground) and the seat back to be adjustable in height (also relative to the ground) and tilt. Provided the chair design meets these requirements and allows the user to achieve a comfortable posture, it is not necessary for the height or tilt of the seat back to be adjustable independently of the seat. Automatic back rest adjustments are acceptable if they provide adequate back support.

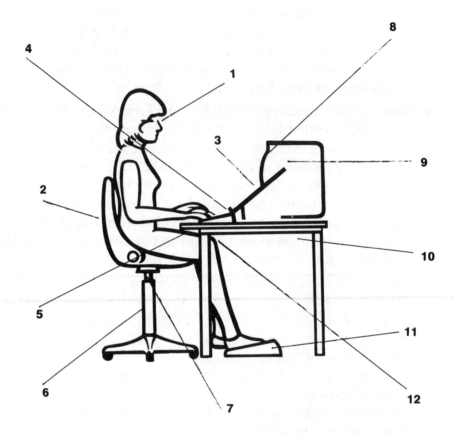

1 Eye height above floor for seated females (0.4m seat) between 1.0 and 1.14m
2 Adjustable backrest for lumbar support – no arm rests
3 Simple document holder
4 Keyboard top approximately 0.7m above floor
5 Accessible work surfaces
6 Swivel chair with stable base, on castors

7 Adjustable seat height approximately 0.4m above floor
8 Screen at approximately right angles to line of sight, but avoiding reflecting light
9 Screen position about 30 to 45° below upright eye height
10 Wires out of way
11 Foot rest for short operators
12 Minimum knee clearance 0.2m between seat and table

● **FIG 8.5 Workstation principles**

Foot rests may be necessary where individual workers are unable to rest their feet flat on the floor (such as when work surfaces cannot be adjusted to the right height in relation to other components of the workstation). Foot rests should not be used when they are not necessary as this can result in poor posture.

Environment

- *Space requirements* Prolonged sitting in a static position can be harmful. It is most important that support surfaces for display screen and other equipment and materials used at the workstation should allow adequate clearance for postural changes. This means adequate clearances for thighs, knees, lower legs and feet under the work surface and between furniture components. The height of the work surface should allow a comfortable position for the arms and wrists, if a keyboard is used.

- *Lighting, reflections and glare* Lighting should be appropriate for all the tasks performed at the workstation, such as reading from the screen, keyboard work, reading printed text, writing on paper and so on. General lighting, by artificial or natural light, or a combination, should illuminate the entire room to an adequate standard. Any supplementary individual lighting provided to cater for personal needs or a particular task should not adversely affect visual conditions at nearby workstations.

- *Illumination* High illumination renders screen characters less easy to see but improves the ease of reading documents. Where a highly illuminated environment is preferred for this or other reasons, the use of positive polarity screens (dark characters on a light background) has advantages as these can be used comfortably at higher levels of illumination than can negative polarity screens.

- *Reflections and glare* Problems that can lead to visual fatigue and stress can arise, for example, from unshielded bright lights or bright areas in the worker's field of view, from an imbalance between brightly and dimly lit parts of the environment, and from reflections on the screen or other parts of the workstation.

Measures that can be taken to minimize these problems include shielding, replacing or repositioning sources of light, rearranging or moving work surfaces, documents all or parts of the workstation, altering the intensity of vertical to horizontal illumination or a combination of these measures. Anti-glare screen filters should be considered as a last resort if other measures fail to solve the problem.

- *Noise* Noise from equipment such as printers at display screen workstations should be kept to levels that do not impair concentration or prevent normal conversation (unless the noise is designed to attract attention, say, to warn of a malfunction). Noise can be reduced by replacement, sound-proofing or repositioning of the equipment. The use of sound-insulating partitions between noisy equipment and the rest of the workstation is an alternative.

- *Heat and humidity* Electronic equipment can be a source of dry heat that can modify the thermal environment at the workstation. Ventilation and humidity should be maintained at levels that prevent discomfort and problems of sore eyes.

Task design and software

- *Principles of task design* Inappropriate task design can be among the causes of stress at work. Stress jeopardizes employee motivation, effectiveness and efficiency and, in some cases, it can lead to significant health problems. The regulations are only applicable where health and safety, rather than productivity, is being put at risk, but employers may find it useful to consider both aspects together as task design changes put into effect for productivity reasons may also benefit health, and vice versa.

  In display screen work, good design of the task can be as important as the correct choice of equipment, furniture and working environment. It is advantageous to:
  - design jobs in a way that offers users variety to exercise discretion, opportunities for learning and appropriate feedback, in preference to simple, repetitive tasks whenever possible (for example, the work of a typist can be made less repetitive and stressful if an element of clerical work is added)
  - match staffing levels to volumes of work, so that individual users are not subject to stress through being either overworked or underworked
  - allow users to participate in the planning, design and implementation of work tasks wherever possible.

- *Principles of software ergonomics* In most display screen work, the software controls both the presentation of information on the screen and the ways in which the worker can manipulate the information. Thus, software design can be an important element of task design. Software that is badly designed or inappropriate for the task will impede the efficient completion of the work and, in some cases, may cause sufficient stress to effect the health of a user. Involving a sample of users in the purchase or design of software can help to provide problems.

Some requirements of the organization and of display screen workers should be established as the basis for designing, selecting and modifying software. In many, though not all, applications, the main points are:

- *suitability for the task* software should enable workers to complete the task efficiently, without presenting unnecessary problems or obstacles
- *ease of use and adaptability* workers should be able to feel that they can master the system and use it effectively, following appropriate training, and the dialogue between the system and the worker should be appropriate to the worker's ability – further, where appropriate, software should enable workers to adapt the user interface to suit their ability level and preferences – and the software should protect workers from the consequences of errors, for example by providing appropriate warnings and information and by enabling 'lost' data to be recovered wherever practicable
- *feedback on system performance* the system should provide appropriate feedback, which may include error messages, suitable assistance ('Help') to workers on request, and messages about changes in the system, such as malfunctions or overloading, and feedback messages should be presented at the right time and in an appropriate style and format – they should not contain unnecessary information
- *format and pace* speed of response to commands and instructions should be appropriate to the task and the workers' abilities and character, cursor movements and position changes should, where possible, be shown on the screen as soon as they are input
- *performance monitoring facilities* quantitative or qualitative checking facilities built into the software can lead to stress if they have adverse results, such as an overemphasis on output speed, but it is possible to design monitoring systems that avoid these drawbacks and provide

information that is helpful to workers as well as managers, although, in all cases, workers should be kept informed about the introduction and operation of such systems.

In the past, display screen work has been responsible for operator stress and ill-health and has attracted much adverse publicity. Attention must be paid to the above factors with a view to eliminating operator stress wherever possible.

ERGONOMICS – THE INTEGRATED APPROACH

The management of human factors is a complicated matter. It requires an integrated approach, covering a broad range of issues – selection, training, supervision, systems and methods of work, workplace design and layout, environmental factors and personal factors.

Ergonomics, fundamentally, is concerned with two issues. First, the promotion of human reliability by removing those aspects of a task, work organization, workplace design and environment predisposed to human error and, second, how people interact with the equipment they operate, that is the man–machine interface.

Factors affecting reliability of performance

There are a number of factors that have a direct effect on human reliability at a given man–machine interface.

- *Work equipment layout* Displays and controls need to be easily distinguished. Displays need to provide accurate, reliable and appropriate information. Controls should be easily operated, with the potential for human error being eliminated or reduced as far as possible.

- *Physical environment* Consideration must be given to environmental stressors, such as temperature, lighting, ventilation, humidity, noise and vibration. In certain cases, appropriate forms of personal protective equipment may be necessary.

- *The system incorporating the man–machine interface* The man–machine interface should be compatible with the total work system. The operator should be fully aware of what is going on around them at all times.

- *The organizational environment* This includes the procedures for recruitment, selection, training and supervision, management styles and supervisor attitudes, availability of rest periods and shift work regimes.

- *Personal factors* People have different attitudes, perceptions, motivators and personalities. They process information at different rates, have differing levels of understanding and perceive risks in different ways. Work does not only create problems for people, they bring their own personal problems to work.

All the above points are significant in the study of human factors (see Figure 8.6 for a summary). Failure by management to consider them is a common cause of those accidents in the workplace that are frequently written down by investigating managers and supervisors to 'carelessness by the operator', 'inattention by the operator' or 'lack of attention'.

CONCLUSION

Ergonomics is a very broad area of study. This study of the needs and requirements of people within a working environment takes in many disciplines. Ergonomics, in its approach to the man–machine interface, has a great contribution to make to the health and safety of people at work. It is concerned with the question of human capability and fallibility and the design of work equipment, in particular, with a view to eliminating the potential for human error. In other words, the basic concept of ergonomics – fitting the task to the person – is a prerequisite for safe systems of work and the elimination of stress in the workplace.

Bearing in mind the requirements of the MHSW to consider human capability when allocating tasks, much greater attention will need to be paid to this subject by management and health and safety practitioners alike.

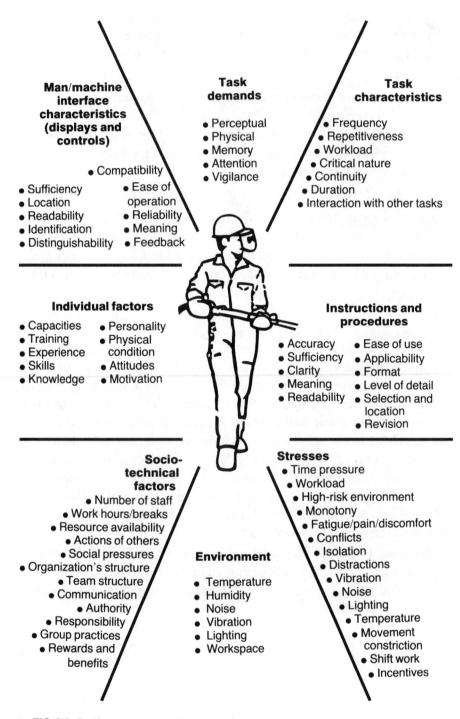

Man/machine interface characteristics (displays and controls)

- Compatibility
- Sufficiency
- Location
- Readability
- Identification
- Distinguishability
- Ease of operation
- Reliability
- Meaning
- Feedback

Task demands

- Perceptual
- Physical
- Memory
- Attention
- Vigilance

Task characteristics

- Frequency
- Repetitiveness
- Workload
- Critical nature
- Continuity
- Duration
- Interaction with other tasks

Individual factors

- Capacities
- Training
- Experience
- Skills
- Knowledge
- Personality
- Physical condition
- Attitudes
- Motivation

Instructions and procedures

- Accuracy
- Sufficiency
- Clarity
- Meaning
- Readability
- Ease of use
- Applicability
- Format
- Level of detail
- Selection and location
- Revision

Socio-technical factors

- Number of staff
- Work hours/breaks
- Resource availability
- Actions of others
- Social pressures
- Organization's structure
- Team structure
- Communication
- Authority
- Responsibility
- Group practices
- Rewards and benefits

Environment

- Temperature
- Humidity
- Noise
- Vibration
- Lighting
- Workspace

Stresses

- Time pressure
- Workload
- High-risk environment
- Monotony
- Fatigue/pain/discomfort
- Conflicts
- Isolation
- Distractions
- Vibration
- Noise
- Lighting
- Temperature
- Movement constriction
- Shift work
- Incentives

● **FIG 8.6 Performance-shaping factors**

Stress at work

WHAT IS STRESS?

Stress is a word that is rarely clearly understood. It means different things to different people. Indeed, almost anything anyone can think of, pleasant or unpleasant, has been described as a source of stress – getting married, being made redundant, getting older, getting a job, too much or too little work, solitary confinement or exposure to excessive noise.

Stress can be defined in many ways, thus:

- 'the common response to attack.' (Hans Selye, 1936)
- any influence that disturbs the natural equilibrium of the living body
- some taxation of the body's resources in order to respond to some environmental circumstance
- the common response to environmental change
- a psychological response that follows failure to cope with problems
- a feeling of sustained anxiety that, over a period of time, leads to disease
- the non-specific response of the body to any demands made on it.

THE COST OF STRESS

Personnel Management in Factsheet 7, July 1988 made the following points:

- stress is said to cost British industry approximately 3 per cent of the gross national product

- stress-related costs amount to more than ten times the cost of all industrial disputes
- stress-related illness directly causes the loss of 40 million working days each year
- the cost of replacing an employee who is underperforming owing to stress is between 50 and 90 per cent of their annual salary
- the total of such replacement costs is about £3 000 000 annually.

THE PHYSIOLOGY OF STRESS

Stress could also be defined simply as the rate of wear and tear on the body systems caused by life. The acknowledged father of stress research, Dr Hans Selye, a Vienna-born endocrinologist of the University of Montreal, in his book *The Stress of Life*, corrected several notions relating to stress, in particular:

- stress is not nervous tension
- stress is not the discharge of hormones from the adrenal glands (the common association of adrenalin with stress is not totally false, but the two are only indirectly associated)
- stress is not simply the influence of some negative occurrence – stress can be caused by quite ordinary, even positive, events, such as a passionate kiss
- stress is not an entirely bad event; we all need a certain amount of stimulation in life and most people can thrive on some forms of stress
- stress does not cause the body's alarm reaction, which is the most common misuse of the expression, what causes the stress reaction is a stressor.

A number of common factors emerge from the definitions of stress earlier and the above comments. Fundamentally, stress is a state manifested by a specific syndrome of biological events. Specific changes occur in the biological system, but they are caused by such a variety of agents that stress is, of necessity, non-specifically induced.

Some stress response, however, will result from any stimulus. Quite simply, a *stressor* produces stress. Stressors may be of an environmental nature, such as extremes of temperature and lighting, noise and vibration (*environmental stressors*). Stress may be induced by isolation, rejection, change within the organization or the feeling that one has been badly treated (*social stressors*). Stress can also be viewed as a general overloading of the body systems (*distress*).

The autonomic system

Stress has a direct association with the autonomic system, which controls an individual's physiological and psychological responses. This is the *flight or fight* system, characterized by two sets of nerves, the sympathetic and parasympathetic, which are responsible for the automatic and unconscious regulation of body function. The *sympathetic system* is concerned with answering the body's call to fight, that is with increased heart rate, more blood to the organs, stimulation of sweat glands and the tiny muscles at the roots of the hairs, dilation of the pupils, suppression of the digestive organs, accompanied by the release of adrenalin and noradrenalin. The *parasympathetic system* is responsible for emotions and protection of the body, which have their physical expression in reflexes, such as widening of the pupils, sweating, quickened pulse, blushing, blanching, digestive disturbance and so on.

The balance between the sympathetic and parasympathetic states is shown in Figure 9.1 and the main signs associated with each state are listed in Table 9.1.

The general adaptation syndrome

Stress is a mobilization of the body's defences – an ancient biochemical survival mechanism perfected during the evolutionary process – allowing human beings to adapt to threatening circumstances.

In 1936, Dr Selye defined this 'general adaptation syndrome', which is comprised of three stages.

1 *The alarm reaction stage* This is typified by receiving a shock at the time when the body's defences are down, followed by a counter-

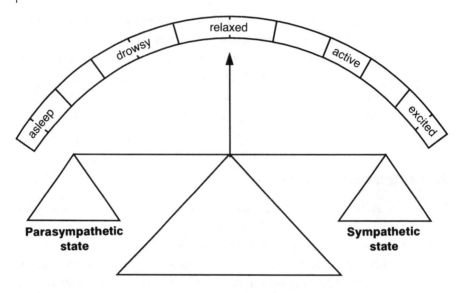

● **FIG 9.1 Sympathetic and parasympathetic balance**

TABLE 9.1 The main signs associated with the parasympathetic and sympathetic states

Parasympathetic state	Sympathetic stage
Eyes closed	Eyes Open
Pupils small	Pupils enlarged
Nasal mucus increased	Nasal mucus decreased
Saliva produced	Dry Mouth
Breathing, slow	Breathing, rapid
Heart rate, slow	Heart rate, rapid
Heart output decreased	Heart output increased
Surface blood vessels dilated	Surface blood vessels constricted
Skin hairs normal	Skin hairs erect (goose pimples)
Dry skin	Sweating
Digestion increased	Digestion slowed
Muscles relaxed	Muscles tense
Slow metabolism	Increased metabolism

shock, when the defences are raised. In physiological terms, once a stressor is recognized, the brain sends forth a biochemical 'messenger' to the pituitary gland that secretes adrenocortitrophic hormone (ACTH). ACTH causes the adrenal glands to secrete corticoids, such as adrenalin. The result is a general 'call to arms' of the body's systems.

2 *The resistance stage* This stage is concerned with two responses: the

body will either resist the stressor or adapt to the effects of the stressor. It is the opposite of the alarm reaction stage, the characteristic physiology of which fades and disperses as the organism adapts to the derangement caused by the stressor.

3 *The exhaustion stage* If the stressor continues to act on the body, however, this acquired adaptation is eventually lost and a state of overloading is reached. The symptoms of the initial alarm reaction stage return and, if the stress is unduly prolonged, the wear and tear will result in damage to a local area or the death of the organism as a whole (see Figure 9.2).

The three stages of the stress response can be summarized as shown in Table 9.2. The effects of stress on the individual are shown in Figure 9.2. This shows the individual surrounded by a variety of stressors. A person's response to these stressors is affected by factors such as their strength of constitution, psychological strength, degree of control over the situation and how they actually perceive the potentially stressful event.

The effect of these stressors is to require some form of general adaptation. Here the situation can go one of two ways. If the individual adapts unsuccessfully, this leads to further wear and tear, weakness and stress-related illness, leading to increased vulnerability to further stressors in their life. Successful adaptation, on the other hand, leads to growth, happiness, security and strength, with greater resistance to further stressors.

THE CAUSE OF STRESS

The causes of stress are diverse, but they can be broadly classified thus (see also Figure 9.3):

- *The physical environmental* Poor working conditions associated with:
- insufficient space
- lack of privacy
- open plan office layouts
- inhuman workplace layouts
- inadequate temperature control so the workplace is too hot or too cold

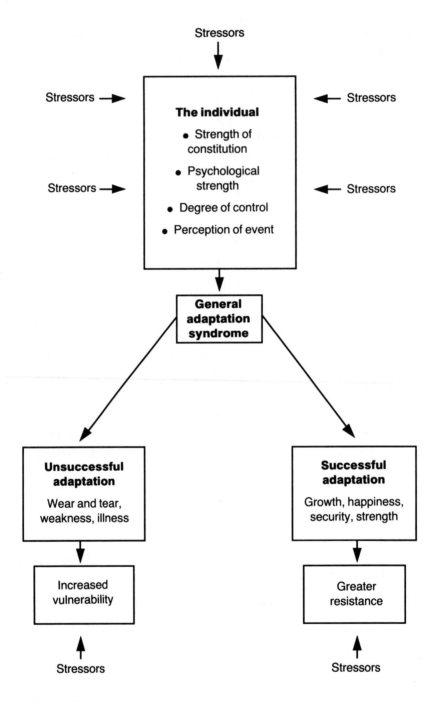

● **FIG 9.2 The General Adaptation Syndrome** (Dr Hans Selye, *The Stress of Life,* McGraw-Hill, New York, 1936)

TABLE 9.2 The three stages of the stress response

The response	What happens	The effect
Fight or flight	Red alert Body and brain prepare for action; extra energy released	Response to danger, meet it and return equilibrium
Secondary	Fats, sugars and cortico-steroids released for more energy	Unless extra fats, etc. used up, then third stage moved into
Exhaustion	Energy stores used up	Serious illness leading to death

- New work patterns
- New technology
- Promotion
- Relocation
- Deregulation
- Downsizing
- Job design
- Boredom
- Noise
- Temperature
- Increased competition
- Longer hours
- Redundancy
- Early retirement
- Acquisition
- Merger
- Manning levels
- Insecurity
- Lighting
- Atmosphere/ventilation

9

- **FIG 9.3 The more common occupational stressors**

- poor levels of illumination
- excessive noise levels
- inadequate ventilation

all are frequent sources of stress in the workplace

- *The organization* The organization, its policies and procedures, culture and style of operation can be a cause of stress due to, for instance:
- insufficient staff for the size of the workload
- too many unfilled posts
- poor coordination between departments
- insufficient training to do the job well
- inadequate information
- no control over the workload

- rigid working procedures
- no time being given to adjust to change

- *The way the organization is managed* Management styles, philosophies, work systems, approaches and objectives can contribute to the individual stress on employees, as a result of:
- inconsistency in style and approach
- emphasis on competitiveness
- 'crisis management' all the time
- information being seen as power
- procedures always being changed
- overdependence on overtime working
- shift work

- *Role in the organization* Everyone has a role within the organization, and stress can result from:
- role ambiguity
- role conflict
- too little responsibility
- lack of managerial support
- responsibility for people and things

- *Relations within the organization* How people relate to each other within the organizational framework and structure can be a significant cause of stress, due to, perhaps:
- poor relations with the boss
- poor relations with colleagues and subordinates
- difficulties in delegating responsibility
- personality conflicts
- no feedback from colleagues or management

- *Career development* Stress is directly related to progression or otherwise in a career and may be created by:
- lack of job security
- overpromotion
- underpromotion
- thwarted ambition
- the job having insufficient status
- not being paid as well as others who do similar jobs

- *Personal and social relationships* The relationships that exist between people on a personal and social basis are frequently a cause of stress due to, for instance:

- not enough opportunities for social contact while at work
- sexism and sexual harassment
- racism and racial harassment
- conflicts with family demands
- divided loyalties between one's own needs and organisational demands

- *Equipment* Inadequate, out-of-date, unreliable work equipment is frequently associated with stressful conditions among workers, as such equipment may be:
- unsuitable for the job or environment
- old and/or in poor condition
- unreliable/not maintained regularly/constantly breaking down
- badly sited
- requires the individual to adopt a fixed and uncomfortable posture
- adds to noise and heat levels

- *Individual concerns* All people are different in terms of attitudes, personality, motivation and in their ability to cope with stressors, so people may experience a stress response due to:
- difficulty in coping with change
- lacking confidence in dealing with interpersonal problems
- not being assertive enough
- not being good at managing time
- lacking knowledge about managing stress.

STRESS WITHIN THE ORGANIZATION

Studies by C. L. Cooper and J. Marshall (1978) into stress within the organization identified an 'organizational boundary', with the individual manager straddling that boundary and, in effect, endeavouring to cope with conflicting stressors created by external demands (the family) and internal demands (the organization) (see Figure 9.4).

The manager's response may be affected by individual personality traits, tolerance for ambiguity, ability to cope with change, specific motivational factors and well-established behavioural patterns.

Within the organization, a number of stressors can be present. These include those associated with:

- *the job* such as too much or two little work, poor physical working conditions, time pressures and the extent and difficulty of the decision-making process
- *role in the organization* for example role conflict, role overload, role ambiguity, specific responsibility for people, lack of participation in the organization's decision-making process
- *career development* for example, the stress associated with overpromotion or underpromotion, lack of job security, thwarted ambitions
- *organizational structure and climate* such as, lack of effective consultation, restrictions on behaviour, office politics
- *relationships within the organization* for example, poor relationship with the boss, colleagues and subordinates, difficulties in delegating responsibility.

On the other side of the organizational boundary is the organization's interface with the outside world. Here, conflict can be created where there may be competition for a manager's time between the organization and family, or between the organization and particular outside interests.

The conclusion drawn from these studies is that organizations should pay attention to the potentially stressful effects of their decisions, management style, consultative arrangements, environmental levels and other matters that can have repercussions for their people and their home lives. The resulting stress can have adverse effects on performance.

Transactional model of stress

This model (see Figure 9.5) depicts:

- *sources of stress at work* that is, factors intrinsic to the job, the individual's role in the organization, career development, relationships at work and the organizational structure and climate
- *individual characteristics* that is, the individual and the effects of the home–work interface as a source of stress
- *symptoms of occupational ill-health* that is, the individual symptoms,

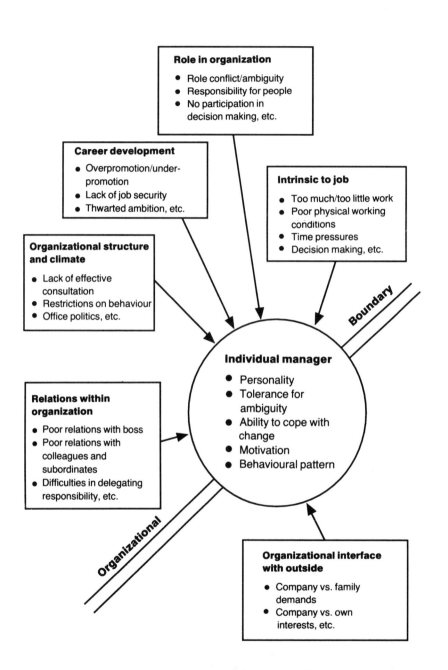

Role in organization

- Role conflict/ambiguity
- Responsibility for people
- No participation in decision making, etc.

Career development

- Overpromotion/under-promotion
- Lack of job security
- Thwarted ambition, etc.

Intrinsic to job

- Too much/too little work
- Poor physical working conditions
- Time pressures
- Decision making, etc.

Organizational structure and climate

- Lack of effective consultation
- Restrictions on behaviour
- Office politics, etc.

Boundary

Individual manager

- Personality
- Tolerance for ambiguity
- Ability to cope with change
- Motivation
- Behavioural pattern

Relations within organization

- Poor relations with boss
- Poor relations with colleagues and subordinates
- Difficulties in delegating responsibility, etc.

Organizational

Organizational interface with outside

- Company vs. family demands
- Company vs. own interests, etc.

• **FIG 9.4 Organizational stresses for the individual manager** (C. L. Cooper and J. Marshall, 1978) such as depression, and the organizational symptoms, such as high labour turnover and high absenteeism

- *the diseases* that can result, not only in terms of coronary heart disease and mental ill-health, but also as far as the organization is concerned, such as prolonged strikes, frequent and severe accidents and chronically poor performance.

Clearly, the source of stress at work must be tackled by the organization before the symptoms manifest themselves.

RESPONSES TO STRESS

Stress can produce both short- and long-term responses. These are summarized in Figure 9.6.

THE EFFECTS OF STRESS ON JOB PERFORMANCE

- *absenteeism* absenteeism, especially on Monday mornings, or in the taking of early/extended meal breaks is a typical manifestation of stress
- *accidents* problem drinkers have three times the average number of accidents (many accidents incorporate stress-related indirect causes)
- *erratic job performance* alternating between low and high productivity due, in some cases, to changes outside the control of the individual, is a common symptom of stress within an organization
- *loss of concentration* stressful events in people's lives commonly result in a lack of the ability to concentrate, whereby a person is easily distracted, or an inability to complete one task at a time
- *loss of short-term memory* this leads to arguments about who said, did or decided what
- *mistakes* stress is a classic cause of errors of judgement, which can result in accidents, wastage, rejects; such mistakes are frequently blamed on others
- *personal appearance* becoming abnormally untidy, perhaps smelling of alcohol, is a common manifestation of a stressful state
- *poor staff relations* people going through a period of stress frequently

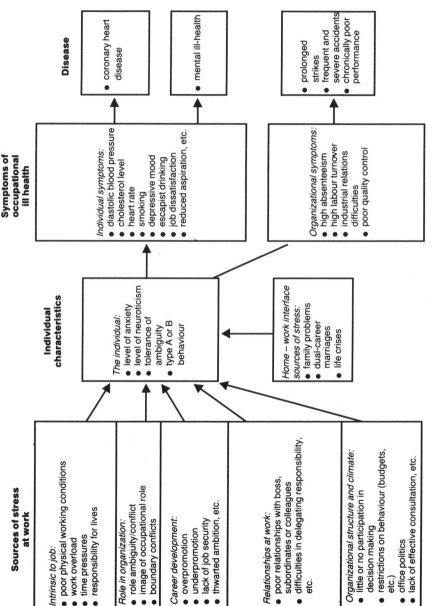

● **FIG 9.5 A model of stress at work**

Short-term responses

Physiological:

- headaches, migraine
- backaches
- eye and vision problems
- allergic skin responses
- disturbed sleep patterns
- digestive disorders

- raised heart rate
- raised blood cholesterol
- raised adrenalin/noradrenalin levels

Mental, emotional, behavioural and social:

- Job dissatisfaction
- anxiety
- depression
- irritability
- frustration
- breakdown of relationships at home and work
- alcohol and drug misuse
- tobacco smoking
- inability to unwind/relax

Long-term responses

Individual health:

- gastric/peptic ulcers
- asthma
- diabetes
- arthritis
- stroke
- high blood pressure
- coronary heart disease
- mental ill-health

Organizational health:

- absenteeism
- poor timekeeping
- high labour turnover
- high sickness absence rates
- low productivity
- industrial unrest

● **FIG 9.6 Responses to stress**

become irritable and sensitive to criticism and this may be accompanied by 'Jekyll and Hyde' mood changes, all of which have a direct effect on staff relationships.

See also Figure 9.7.

ANXIETY AND DEPRESSION

Anxiety and depression are the classic manifestations of stress.

- *Anxiety* This is a state of tension coupled with apprehension, worry, guilt, insecurity and a constant need for reassurance. It is accompanied by a number of psychosomatic symptoms, such as profuse perspiration, difficulty in breathing, gastric disturbances, rapid heartbeat, frequent urination, muscle tension or high blood pressure. Insomnia is a reliable indicator of a state of anxiety.

How you feel
You are anxious, aggressive, apathetic, bored, tired, depressed, frustrated, guilty, irritable, lacking in confidence, tense, nervous and lonely.

How you behave
You have accidents, take drugs, get emotional, eat too much or too little, drink and smoke excessively, have incoherent speech, nervous laughter, are restless, trembling.

How you think
You have difficulty in making decisions, are less creative in solving problems, forgetful, hypersensitive to criticism and have poor concentration and poor work organization and completion of tasks.

What happens to your body
You may experience increased heart rate and blood pressure, dryness of the mouth, sweating, pupil dilation, hot and cold spells, 'a lump in the throat', numbness, butterflies in the stomach.

What happens to your health
You might suffer from asthma, chest and back pains, coronary heart disease, faintness and dizziness, dyspepsia, frequent urination, headaches and migraine, neuroses, nightmares insomnia, psychoses, skin complaints, ulcers and loss of sexual interest.

How it affects your work
The results could include increased absenteeism, poorer communications and industrial relations, less commitment, a higher accident rate, more antagonism, less creativity, less concern for your fellow workers, less job satisfaction, poorer productivity.

● **FIG 9.7 What stress can do to you**
   (Adapted from T. Cox, *Stress*, Macmillan, 1978)

● *Depression* Depression, on the other hand, is much more a mood, characterized by feelings of dejection and gloom, and other permutations, such as feelings of hopelessness, futility and guilt. The well-known American psychiatrist David Viscott, described depression as 'a sadness which has lost its relationship to the logical progression of events'. It may be mild or severe. Its milder form may be a direct result of a crisis in work relationships. Severe forms may exhibit biochemical disturbances, and the extreme form may lead to suicide.

ROLE THEORY

Role Theory sees large organizations as systems of interlocking roles. These roles relate to what people do and what other people expect of them rather than their individual identities.

An individual's thoughts and actions are influenced by identification with that role. Everyone in a role has contact with people – superiors, subordinates, external contacts or contractors – who communicate their expectations of the role holder, trying to influence their behaviour and subjecting them to feedback. The individual, therefore, has certain expectations about how people should behave according to their status, age, function and responsibility. These expectations form the basis for a standard by which individual behaviour is evaluated, as well as a guide for reward.

Stress arises in this framework due to role ambiguity, role conflict and role overload.

- *Role ambiguity* This is the situation in which the role holder has insufficient information to adequately perform their role or the information is open to more than one interpretation. Potentially ambiguous situations occur in jobs where there is a time lag between the action taken and visible results – or where the role holder is unable to see the results of their actions.

- *Role conflict* Role conflict arises where members of the organization, who exchange information with the role holder, have different expectations of their roles. Each may exert pressure on the role holder. Satisfying one expectation could make compliance with other expectations difficult. This is the classic 'servant of two masters' situation.

- *Role overload and role underload* Role overload results from a combination of role ambiguity and role conflict. The role holder works harder to clarify normal expectations or to satisfy conflicting priorities that are impossible to achieve within the time limits specified. Similarly, certain people, who may have had a demanding job, may be shunted into a job where there is too much time available to complete an identified workload, resulting in boredom, excessive attention to minute details, as far as subordinates are concerned, and a general feeling of isolation. Role underload can be as significant a cause of stress as role overload.

Research has shown that where experience of role ambiguity, conflict and overload/underload is high, then job satisfaction is low. This may well be coupled with anxiety and depression, factors that may add to the onset of stress-related conditions, such as peptic ulcers, coronary heart disease and nervous breakdowns.

PERSONALITY AND STRESS

Personality was defined in Chapter 4 as 'the dynamic organization within the individual of the psychophysical systems that determine his characteristic behaviour and thought' (G. W. Allport, 1961).

Personality traits and how they relate to stress

Various types and traits of personality have been established by psychological researchers over the last 30 years. These can be classified as follows:

- *type A, ambitious* active and energetic, impatient if they have to wait in a queue, conscientious, maintain high standards, time is a problem – there is never enough – often intolerant of others who may be slower

- *type B, placid* quiet, very little worries them, often uncompetitive, put their worries into things they can alter and leave others to worry about the rest

- *type C, worrying* nervous, highly strung, not very confident of self-ability, anxious about the future and of being able to cope

- *type D, carefree* love variety, often athletic and daring, very little worries them, not concerned about the future

- *type E, suspicious* dedicated and serious, very concerned with others' opinions of them, do not take criticisms kindly and remember such criticisms for a long time, distrust most people

- *type F, dependent* bored with their own company, sensitive to surroundings, rely on others a great deal, people who interest them most are oddly unreliable, they find the people that they really need to be boring, do not respond easily to change

- *type G, fussy* punctilious, conscientious and like a set routine, do not like change, any new problems throw them because there are no rules to follow, conventional and predictable, collect stamps and coins and keep them in a beautifully ordered state, great believers in authority.

Research indicates that most people combine traits of more than one of these types and so the definitions above can only be used as a guide. The type most at risk to stress is type A.

CRISIS

The word 'crisis' can be defined as 'a situation when something happens that requires major decisions to be made quickly'. No two people necessarily respond to a perceived crisis in the same way, however. Similarly, a crisis situation to one person may be something relatively minor to another.

Crisis situations

Typical crisis situations could be associated with sudden bad health, a client going bankrupt, loss of money, moving house at very short notice, family difficulties, being made redundant or losing important documents, such as a cheque book, diary or credit cards.

Ideally, we need to identify crises before they arise and have contingency plans ready to cope, for instance having colleagues to stand in and cover the job in the event of unexpected illness, having alternative sources of finance available, retaining extra copies of important documents. Most crisis situations are, however, totally unexpected.

The process of transition (Wei-chi)

'Wei-chi' is Chinese for crisis. The more literal transition is 'from danger comes opportunity', indicating a slow process of transition, the various stages of which can be stressful. These various stages (shown in Figure 9.8) are as follows:

1 denial, shock and a feeling of numbness when presented with an unexpected adverse situation, such as a financial loss or a notice of redundancy
2 euphoria–endeavouring to minimize the effects of the shock
3 searching and questioning oneself as to what went wrong, accompanied by pining for a return to the 'status quo'
4 anger
5 guilt

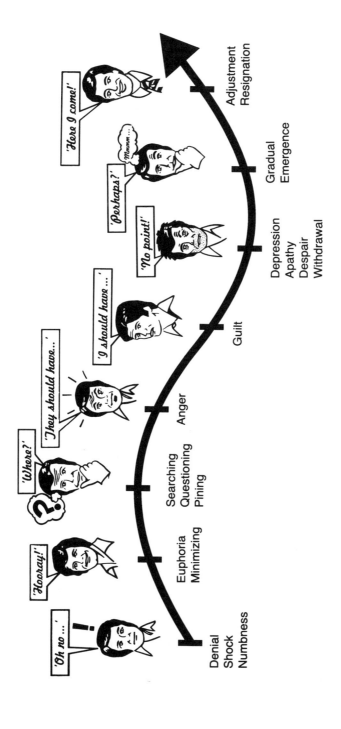

● **FIG 9.8** *Wei-chi* – the process of transition

6 depression, apathy, despair, withdrawal

7 gradual emergence from the stressful and depressive state

8 adjustment, resignation, rebuilding, new ideas – 'a chance to do what I've always wanted to do', success.

Crisis management has much to offer both the individual and the organization. It is important that people should understand and recognize these various stages with a view to designing strategies and systems to prevent their recurrence.

SMOKING, ALCOHOL AND DRUGS

Stress can result in increased addiction to smoking and alcohol consumption and, occasionally, drug addiction. While the adverse health effects of smoking are well-known, alcohol and drug addiction need some consideration.

Alcohol addiction

Alcoholism is a true addiction and the alcoholic should be encouraged to obtain medical help and advice. Alcohol abuse leads to broken homes, broken marriages, lost jobs, a certain amount of crime and unhappiness generally for those who may come into contact with the alcoholic, together with varying degrees of physical and mental disease. On the other hand, it is a fact that many people consume very large quantities of alcohol throughout a long life without showing any apparent ill-effects whatever and that, in most cases, alcoholism is a symptom rather than a disease.

The general, although by no means universally accepted, belief today is that the physical diseases brought about by the excessive consumption of alcohol are the result of its indirect effect of producing malnutrition rather than its direct toxic one. The repeated consumption of strong spirits, especially on an empty stomach, can lead to chronic gastritis, and possible inflammation of the intestines that interferes with the absorption of food substances, notably those in the vitamin B group. This, in turn, damages the nerve cells, causing alcoholic neuritis, injury to the brain cells leading to certain forms of insanity and, in some cases cirrhosis of the liver.

The alcoholic is not necessarily the person who becomes obviously drunk, frequently, but, more commonly, the person who drinks steadily throughout the day, often without there being any immediate effect that is apparent to others. Later, however, symptoms that are partly due to physical effects, partly to the underlying neurosis at the root of the trouble, in most cases, and partly social, begin to show themselves. The individual eats less and drinks more, often begins the day with vomiting or nausea, which necessitates taking the first drink before they can face the public, their appearance tends to become bloated and the eyes are often red and congested. Their work suffers, they forget to keep appointments and become indifferent to their social responsibilities. Their craving for drink becomes insatiable and, when they are unable to get it, they become shaky, irritable and tense. As they are ashamed of their condition, they try to hide it and, often, instead of drinking openly, hide the bottles about the house and, perhaps, in the office. Their emotions are less controlled and they get angry or tearful readily, tell their facile lies and a minor illness, or going without a drink for a time, may lead to an attack of the 'DTs' (delirium tremens).

In severe cases, the alcoholic may die from cirrhosis of the liver or an attack of pneumonia or some other infection, that would not generally be fatal to healthy people, may be so in their case. No matter how alcoholism manifests itself, the alcoholic needs help, particularly if the condition is prejudicing the safety of their fellow workers. In most cases, this implies complete abstention for a period of time under controlled conditions, away from the normal temptations of the home and the workplace, perhaps psychotherapy, to assess any psychological causes of the condition, and the general building up of impaired physical health.

It is in cases of alcoholism that the occupational health practitioner (occupational physician and occupational health nurse) can be of considerable support and assistance in bringing about the gradual rehabilitation necessary, perhaps by advising on the various social and therapeutic treatments available. The occupational health practitioner is also trained in the early detection of cases of alcohol abuse and, through counselling and routine surveillance, can prevent the situation from deteriorating further.

Figure 9.9 is a useful guide for managers as to the kind of evidence of alcohol addiction that may be observed among staff. It shows the various phases in the development of alcohol addiction, the level of efficiency at these phases, various crisis points during the gradual deterioration of the individual and the visible signs during the various phases.

Statements of company policy

Many organizations have, over the last decade, produced statements of company policy on both smoking at work and alcohol at work, with a view to improving standards of employee health and reducing the sickness absence associated with smoking and alcohol consumption in the workplace and during working hours.

Drugs and drug addiction

The expression 'drug' is popularly associated with a narcotic or habit-forming substance. Technically, however, a drug is a substance taken medicinally to assist recovery from sickness, or to relieve symptoms of a disease or condition, or to modify any natural body process. However, many people see drug-taking as the panacea for stress, relying on tranquilizers to reduce anxiety and amphetamines (pep pills) to counter fatigue. Such drug-taking represents a major health risk, particularly if the individual consumes alcohol at the same time. In some cases, the consumption of prescribed drugs can be a contributory factor in accidents.

In the last decade the threat of the elicit use of drugs, and their effects on society generally, has increased dramatically. In considering the problem of drugs and drug addiction, it is necessary to distinguish, therefore, between the risks to health and safety of people who may be taking drugs prescribed by a registered medical practitioner and those who are engaging in illegal drug-taking. Both groups may represent a significant risk to the health and safety of themselves and other persons who may be affected by their activities at work.

Whatever the case, it is essential that local management are aware of the situation respecting individual employees. In the case of people who are taking prescribed drugs, the view of an occupational physician or occupational health nurse may be necessary before employment in a particular activity is permitted. Where employees, on the other hand, exhibit evidence of the elicit use of drugs or their effects

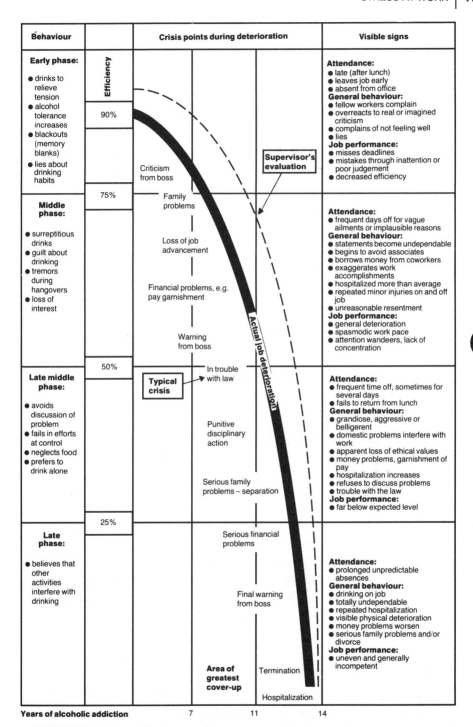

Behaviour		Crisis points during deterioration		Visible signs

Early phase:

- drinks to relieve tension
- alcohol tolerance increases
- blackouts (memory blanks)
- lies about drinking habits

Efficiency — 90%

Criticism from boss

Supervisor's evaluation

75%

Middle phase:

- surreptitious drinks
- guilt about drinking
- tremors during hangovers
- loss of interest

Family problems

Loss of job advancement

Financial problems, e.g. pay garnishment

Warning from boss

Actual job deterioration

50%

Late middle phase:

- avoids discussion of problem
- fails in efforts at control
- neglects food
- prefers to drink alone

Typical crisis → In trouble with law

Punitive disciplinary action

Serious family problems – separation

25%

Late phase:

- believes that other activities interfere with drinking

Serious financial problems

Final warning from boss

Area of greatest cover-up

Termination

Hospitalization

Visible signs:

Attendance:
- late (after lunch)
- leaves job early
- absent from office

General behaviour:
- fellow workers complain
- overreacts to real or imagined criticism
- complains of not feeling well
- lies

Job performance:
- misses deadlines
- mistakes through inattention or poor judgement
- decreased efficiency

Attendance:
- frequent days off for vague ailments or implausible reasons

General behaviour:
- statements become undependable
- begins to avoid associates
- borrows money from coworkers
- exaggerates work accomplishments
- hospitalized more than average
- repeated minor injuries on and off job
- unreasonable resentment

Job performance:
- general deterioration
- spasmodic work pace
- attention wanders, lack of concentration

Attendance:
- frequent time off, sometimes for several days
- fails to return from lunch

General behaviour:
- grandiose, aggressive or belligerent
- domestic problems interfere with work
- apparent loss of ethical values
- money problems, garnishment of pay
- hospitalization increases
- refuses to discuss problems
- trouble with the law

Job performance:
- far below expected level

Attendance:
- prolonged unpredictable absences

General behaviour:
- drinking on job
- totally undependable
- repeated hospitalization
- visible physical deterioration
- money problems worsen
- serious family problems and/or divorce

Job performance:
- uneven and generally incompetent

9

Years of alcoholic addiction   7   11   14

● **FIG 9.9 The experience of the alcoholic**

whilst at work, police involvement is necessary.

WOMEN AT WORK

Research has shown that women can be subject to many stressors at work that are not suffered by their male counterparts. While sexual harassment at work is a common cause of stress among women, other causes include:

- performance-related pressures
- lower rates of pay
- the problem of maintaining dependents at home
- lack of encouragement from superiors, including not being taken seriously
- discrimination in terms of advancement
- sex discrimination and prejudice
- pressure from dependents at home
- career-related dilemmas, including whether or not to start a family or marry or live with someone
- lack of social support from colleagues
- lack of same-sex role models
- evidence of male colleagues being treated more favourably by management
- being single and labelled as an oddity
- lack of domestic support at home.

Management should be aware of the various forms of stress to which women are exposed at work and take measures, including disciplinary measures, where evidence of, particularly, sexual harassment exists.

SHIFT WORK

Current facts

Approximately 29 per cent of employees in the UK work some form of shift pattern, and 25 per cent of employees undertake night shifts. Researchers have, over the years, studied the physical and psychological effects of atypical working on these groups of people, in particular factory workers and transport workers, and have reported a number of findings. For instance:

- 60 to 80 per cent of all shift workers experience long-standing sleeping problems
- shift workers are 5 to 15 times more likely to experience mood disorders as a result of poor-quality sleep
- drug and alcohol abuse are much higher among shift workers
- 80 per cent of all shift workers complain of chronic fatigue
- approximately 75 per cent of shift workers feel isolated from family and friends
- digestive disorders are four to five times more likely to occur among shift workers
- from a safety viewpoint, more serious errors and accidents, resulting from human error, occur during shift work operations.

9

The appeal of shift work

So, why are people prepared to do shift work? In some manufacturing activities, there may be no alternative to shifts, due to the need, principally, to keep the production process running continuously. However, because of the disturbing effects of shift work, it has always tended to attract higher pay than standard day work – in some cases, a shift work premium. Such a premium frequently appeals to younger workers who may need the extra money at a particular point in their lives, for example, to raise the deposit for a house, but who intend, later, once this financial objective has been achieved, to revert back to routine day shifts.

The problem is, of course, that once people have become used to the improved pay and accompanying lifestyle, they tend to get locked into the system of shift work. What they do not appreciate is the stress that this can create in their own lives, in the live s of their family and friends, and the changes in health state that can take place gradually over a period of years.

Shift work and stress

Fundamentally, very few people are trained to appreciate the stress that shift work can create and the strategies that are necessary in order to cope with it. Further, they do not recognize that stress is all about how well or badly people adapt to changes in their lives, in particular the gradual adaptation process that must take place when exposed to this type of stress if they are to survive (as we saw in Dr Hans Selye's General Adaptation Syndrome earlier in this chapter). They have probably never been advised, for instance, that adjustment to shift work is very much age-related. Younger people can adapt to the way of life better, but as people get older sleep patterns become more important and so older shift workers frequently find that they may be lacking sleep or that it is frequently broken. This is a particular cause of stress.

In effect, they become part of a subculture (often encountered among permanent night workers). Because they are asleep while everything within the family group is taking place, they totally lose touch with members of the family and friends. This can produce a feeling of guilt, frustration and of isolation from society, resulting in changes in attitude, short-temperedness, personality change and loss of motivation.

Conflict can arise – particularly if the family is unaware of the stress associated with shift work – between the personal needs of the shift worker and social and family needs, particularly one's responsibilities as a father or mother. Inevitably, many people try to get the best of both worlds by leading a normal everyday life during the day and work shifts at the same time. This changing of work and sleep schedules has a profound effect on their body clocks, resulting in significant physical and emotional problems over a period of time.

The principal causes of stress

The psychological factors that affect an individual's ability to make the adjustments required by various work schedules are:

- *age* young people adapt more easily to changing shift patterns as their need for sleep is less than that of an older person

- *sleep needs* some people need less sleep than others and can adjust to shift work more easily as a result, but sleep and sleep patterns are a complicated psychological phenomenon – sleep takes place in a series of four stages, and because of daytime disturbances, many shift workers do not experience the beneficial delta, or deep sleep, stage, so important in terms of physical restoration

- *sex* women can experience complicated problems in adapting to shift work, particularly in their reproductive cycles.

- *'day people' and 'night people'* some people are more naturally alert in the morning and so find it more difficult to adjust to shift changes; other people perform more efficiently at night – much of this variation is associated with arousal levels, which vary from person to person

- *the type of work* generally, people experience less stress where undertaking work that requires physical activity, such as assembly work, than work which is inactive because sleep-deprived people are likely to lose concentration, particularly if undertaking work of a monitoring nature, such as inspection work

- *desynchronization of body rhythms* long-term problems arise when people rotate schedules at a rate that is too fast for their body to adjust to as body rhythms get out of synchrony with the external environment and internal processes, normally synchronized, such as the digestive system, begin to drift apart. (This drifting happens because internal processes adjust at different rates. It leads to longstanding tiredness, a feeling of being generally run down, depression – a classic manifestation of stress – and lack of energy.)

The health and safety implications

In many organizations, accident rates are substantially higher for shift and night work operations. Many people would write this fact down to reduced supervision levels, a lack of training of shift-workers in safe working practices or to the view that shift workers 'couldn't care less'. Very rarely do senior management endeavour to ascertain the reasons for these high accident rates, however.

Sleep deprivation results in chronic fatigue in a substantial number of shift workers – 60 to 80 per cent, in fact. Fatigue is frequently associated with impaired memory, judgement, reaction time and concentration. It is not uncommon for shift workers to doze off on the job or even be found asleep 'Falling asleep at the wheel', especially between the hours of 1 and 4 a.m., is a common problem with HGV drivers. A Japanese study showed that 82 per cent of near miss train accidents occurred between midnight and 8 a.m. Also the Exxon Valdez ran aground a few minutes after midnight, due to the crew having been working 'heavy overtime'. Many other major incidents, such as those at Three Mile Island, Bhopal and the Challenger explosion also occurred between midnight and 8 a.m. Most of the reports on these incidents indicate 'human error due to operator fatigue' as a significant contributory cause of the accident concerned.

Things the individual can do to reduce the stress of shift work

Strategies are available that are aimed at minimizing the desynchronization of body rhythms and other health problems associated with shift working. The principal objective is to stabilize body rhythms and to provide consistent time cues to the body.

First, there is a need to recognize that workers must be trained to appreciate the potentially stressful affects of shift working and that there is no perfect solution to this problem. However, they do have some control over how they adjust their lives to the working arrangements and the change in lifestyle that this implies.

Second, they need to plan their sleeping, family and social contact schedules in such a way that the stress of this adjustment is minimized. Most health problems arise as a result of changing daily schedules at a rate quicker than that at which the body can adjust. This can result, as we have seen, in desynchronization, with reduced efficiency, generally due to sleep deprivation.

Other factors need to be considered, however, particularly the following.

- *Sleep deprivation* This can have long-term effects on the health of the shift worker. It is important to consider individual lifestyle – in particular, diet – the actual environment in which sleep takes place, family and social relationships and, in certain cases, the use of alcohol and drugs.

- *Diet* A sensible dietary regime, taking account of the difference between the time of eating and the timing of the digestive system, will assist the worker to minimize discomfort and digestive disorders.

- *Alcohol and drugs* Avoidance of alcohol and drugs, including caffeine and nicotine, can result in improved sleep quality. The occasional use of sleeping tablets may be beneficial, but should be used under medical supervision.

- *Family and friends* They should appreciate the demands on the shift worker and make every effort to assist in reducing the potentially stressful effects of this type of work. In particular, better planning of family and social events is necessary to reduce the isolation frequently experienced by shift workers.

Things the organization can do to reduce the stress of shift work

Shift work has been a feature of British industry for over a century. For some people, it is a way of life that they have adjusted to with ease over a period of time. For others, it can be significantly stressful, resulting in a wide range of psychological and health-related symptoms and effects, as we have seen, together with an increased potential for accidents. However, a number of remedies are available to organizations. These include:

- consultation prior to the introduction of shift work
- recognition by management that shift work can be stressful for certain groups of workers and of the need to assist in their adjustment to this type of work
- regular health surveillance of shift workers to identify any health deterioration or change at an early stage
- training of shift workers so they can recognize the potentially stressful effects and the changes in lifestyle that may be needed to reduce these stressful effects
- better communication between management and shift workers to reduce the feeling of isolation frequently encountered among such workers.

THE EVALUATION OF STRESS

Major changes in people's lives, such as marital separation, changes in responsibility at work, job loss and even getting married, can be stressful.

Research by Drs D. Holmes and J. Rahe of the School of Medicine, University of Washington, USA, into the clinical effects of major life changes has identified the concept of the *life change unit* (LCU). This is a unit of individual stress measurement that can be related to the impact expected on the person's health. Over a period of 20 years, Drs Holmes and Rahe were able to assign a numerical value to a range of *life events*, such as a son or daughter leaving home, change in residence or death of a family member, and rank them according to their magnitude and importance. Then they compared the LCU scores of some 5000 individuals with their respective medical histories. They concluded that those with a high rating on the *life change index* were more likely to become ill.

Their *Schedule of Recent Events* based on the total number of LCUs experienced in a year – has since been applied to many groups, confirming the view that the higher the degree of life change within a period of time, the greater the risk of subsequent illness, regardless of whether or not the change is perceived as desirable or undesirable.

According to Drs Holmes and Rahe, if an individual's LCUs total 150 to 199, they stand a mild chance of illness in the following year, while a total of 200 to 299 creates a moderate risk. Over 300 LCUs puts them in the group very likely to suffer serious physical or emotional illness.

The lessons to be learned from this theory are that people should try to regulate the changes in their lives, most of which are under their control, and endeavour to stagger their incidence and intensity. Table 9.3 shows life change events and their LCU ratings. The values are, of course, averages.

COPING WITH STRESS

Some ideas for responding to stress.

The following is adapted from Adams, 1980. There are a number of

TABLE 9.3 Drs Holmes and Rahe's scale of life change units

Event	LCUs	Event	LCUs
Death of a spouse	100	Change in work responsibilities	30
Marital separation	65	Son/daughter leaving home	29
Death of close family member	63	Trouble with in-laws	29
Personal injury or illness	53	Outstanding personal achievement	29
Marriage	50	Wife beginning or stopping work	29
Loss of job	47	Revision of personal habits	24
Marital reconciliation	45	Trouble with business superior	23
Retirement	45	Change in work hours or conditions	20
Change in health of a family		Change in residence	20
member	44	Change in schools	20
Wife's preganancy	40	Change in recreation	19
Sexual difficulties	39	Change in social activities	18
Gain of a new family member	39	Taking out a small mortgage	17
Change in financial status	38	Change in sleeping habits	16
Death of a close friend	37	Change in number of family	
Change to a different kind of work	36	get-togethers	15
Increase or decrease in arguments		Change in eating habits	15
with spouse	35	Holiday	13
Taking out a bigger mortgage	31	Minor violations of the law	11
Foreclosure of mortgage or loan	30		

ways in which one can manage the stress in one's life. We are all unique and what works well for one person may be completely ineffective for another. Here is a range of ideas for responding to stress, each of which has worked well for someone at some time.

● *Become more knowledgeable about stress*
– understand the process and effects of stress
– identify your major source of stress–situations, people and so on
– anticipate stressful periods and plan for them
– develop a repertoire of successful stress management techniques and practise them
– learn to identify the opportunities for personal growth inherent in periods of stress
– find the level of stress that is best for you, remembering that both insufficient and excessive stress are potentially harmful.

● *Take a systematic approach to problem solving*
– define your problem specifically – divide it into manageable components that can be dealt with easily

- gather sufficient information about the problem and put it into perspective
- discover why the problem exists for you
- review your experience with the present problem or similar ones
- develop and evaluate a set of alternative courses of action
- select a course of action an proceed with it.

- *Come to terms with your feelings*
- differentiate between your thoughts and your feelings
- do not suppress your feelings; acknowledge them to yourself, and share them with others
- learn to be flexible and adaptive
- accept your feelings.

- *Develop effective behavioural skills*
- don't use the word 'cannot' when you actually mean 'will not'
- when you have determined what needs to be done with your life, *act on your decisions*
- use free time productively
- be assertive
- manage conflicts openly and directly
- avoid blaming others for situations
- provide positive feedback to others
- learn to say 'No'
- deal with problems as soon as they appear; if you procrastinate, they may intensify
- evaluate the reality of your expectations, avoiding both the grandiose and the catastrophic
- learn to let go of situations and take breaks.

- *Establish and maintain a strong support network*
- ask for direct help and be receptive when it is offered
- develop empathy for others
- make an honest assessment of your needs for support and satisfaction with the support you currently receive
- list six people with whom you would like to improve your relationship and, in each case, identify one action step you will take towards such improvement
- rid yourself of dead and damaging relationships
- maintain high-quality relationships
- tell the members of your support network that you value the relationships shared with them.

- *Develop a lifestyle that will buffer you against the effects of stress*
  - regularly practise some form of each of the following types of exercise – vigorous stretching and recreational
  - engage regularly in some form of systematic relaxation
  - use alcohol in moderation or not at all
  - do not use tobacco
  - obtain sufficient rest on a regular basis
  - eat a balanced diet
  - avoid caffeine
  - avoid foods high in sugar, salt, white flour, saturated fats and chemicals
  - plan your use of time both on a daily and long-term basis
  - seek out variety and change of pace
  - take total responsibility for your life
  - maintain an optimistic attitude
  - do not dwell on unimportant matters

- *Concentrate on positive spiritual development*
  - adopt the attitude that no problem is too monumental to be solved
  - engage regularly in prayer or meditation
  - establish a sense of purpose and relaxation
  - seek spiritual guidance
  - learn to transcend stressful situations
  - believe in yourself
  - increase your awareness of the interdependence of all things in the universe.

- *Plan and execute successful lifestyle changes*
  - expect to succeed
  - approach projects one step at a time
  - keep change projects small and manageable
  - practise each change rigorously for 21 days, then decide whether or not to continue with it
  - celebrate your success; reward yourself.

How to relax

- Choose a quiet place where you will not be disturbed.
- Lie down comfortably and ensure that you are warm enough because, as you relax, your body temperature will fall slightly.

- Close your eyes and take three deep breaths in and sigh them out. This relaxes the diaphragm, and therefore, your breathing.

- Mentally, go through your body, physically tightening and then relaxing each part. Feel for areas of tenseness and then feel them relax on your breaths out.

- Ignore outside noise interruptions by thinking of a lovely colour or a beautiful place or the sound of water. Let your mind and body float. If stray thoughts occur, just let them pass through your mind. If appropriate, relaxation tapes, incense or aromatic oils can all aid relaxation.

- To recover, gradually deepen your breathing, start moving your muscles gently and, as you 'awake', very gently arouse yourself. Get up by rolling on to your side and sitting before standing to avoid dizziness.

- When you are fully awake, stretch and take three deep breaths. After a few minutes you will feel refreshed and really alert. Ideally, follow a relaxation period with a walk in the fresh air.

MANAGING STRESS

Potentially stressful organizations are those:

- that are large and bureaucratic
- in which there are formally prescribed rules and regulations
- where there is conflict between positions and people
- where people are expected to work hard for long hours
- where no praise is given
- where the general culture is classified as 'unfriendly'
- where there is conflict between normal work and outside interests.

Why do something?

- *Cost benefits* There are obvious cost benefits associated with reduced absenteeism and accidents, together with their related direct and indirect health and other costs. Generally, however, an organization will be more effective if there is conscious recognition of stress potential and efforts are made to eliminate or reduce them.

- *Morale* One of the standard criticisms from people at all levels is that the organization does not care about its people. This feeling is reflected in attitudes to management, the job and the organization as a whole. It is important, therefore, for the organization to show that it really does care at all levels. This will result in increased motivation and a genuine desire on the part of staff to perform better. There is clear-cut evidence throughout the world that shows that the most profitable companies are those that take an interest in their staff and promote a caring approach.

What can be done?

There is a need here to consider both organizational and individual strategies for managing stress in the workplace.

9

Organizational strategy

- *Employee health and welfare* Various strategies are available for ensuring the sound health and welfare of employees. These include various forms of health surveillance, health promotion activities, counselling on health-related issues and the provision of good-quality welfare amenity provisions, such as sanitation, washing, showering facilities, facilities for taking meals and so on.

- *Management style* Management is frequently seen as uncaring, hostile, uncommunicative and secretive. A caring philosophy is essential, together with sound communication systems and openness on all issues that affect staff.

- *Change management* Most organizations go through periods of change from time to time. Management should recognize that impending change, in any form, is one of the most significant causes of stress at work. It is commonly associated with job uncertainty, insecurity, the threat of redundancy, the need to acquire new skills and techniques, perhaps at a late stage in life, relocation and loss of promotion prospects. To eliminate the potentially stressful effects of change, a high level of communication in terms of keeping staff informed about what is happening should be maintained and any such changes should be well-managed on a stage-by-stage basis.

- *Specialist activity* Specialist activities, such as those involving the selection and training of staff, should take into account the potential for stress in certain work activities. People should be trained to recognize the stressful elements in their work and the strategies available for coping with these stressors. Moreover, job design and work organization should be based on ergonomic principles.

Individual strategy

There may be a need for individuals to:

- develop new skills for coping with the stress in their lives
- receive support by counselling and other measures
- receive social support
- adopt a healthier lifestyle
- where appropriate, be prescribed drugs for a limited period

The use of occupational health practitioners is recommended in these circumstances.

Stress management action plans

Any action plan to deal with stress at the organizational level should follow a number of clearly defined stages, as follows:

- recognize the causes and symptoms of stress
- decide that the organization needs to do something about it
- decide which group or groups of people can least afford to be stressed, say, key operators, supervisors
- examine and evaluate by interview and/or questionnaire the specific causes of stress
- analyse the problem areas
- decide on suitable strategies, such as counselling, social support, training, such as Time Management, environmental improvement and control, redesign of jobs, ergonomic studies.

At the individual level, people should take the following action:

- identify their work and life objectives and re-evaluate them on a regular basis or as necessary, putting them up where they can be seen

- ensure a correct time balance

- identify their stress indicators and plan how these sources of stress can be eliminated – see them as red stop lights

- allow 30 minutes each day for refreshing and recharging

- identify crisis areas and plan contingency action

- identify key tasks and priorities and do the *important*, not necessarily the *urgent*

- keep their eyes on their objectives and, above all, have fun!

CONCLUSION

Some managers are not prepared to recognize the problem of stress in the workplace. The common response to people complaining of stress is 'If you can't stand the heat, get out of the kitchen!' However, it wasn't until people such as occupational health nurses started to relate sickness absence levels to stress that managers eventually began to admit that the results of their decisions and actions, the environment they provided for operators and many other features of their organizational activities could be stressful. Fortunately, the type of manager who dismisses stress is rapidly disappearing as the recognition of the existence of stress and the human factors-related approach to management become more common.

The problem is still with us, however. Many people simply fail to recognize actual stressful situations or future stressful situations in their lives. These situations can arise from problems at home, in their relationships with people or as a result of a specific life event, such as a bereavement. Generally, they 'bear up' and endeavour to cope. In most cases, they do, but with varying effects on their health, some of which can be serious.

If people are to cope with stressful situations, they have got to go back to the basic principles, namely:

- identify the sorts of events in their lives that create the stress response
- measure and evaluate the significance of these events
- learn various forms of coping strategies to enable them to deal within these life events.

People would be a lot happier if they would undertake this exercise.

10

A health and safety culture

INTRODUCTION

The word 'culture' has been variously defined as:

a state of manners, taste and intellectual development at a time or place
(Collins Gem English Dictionary)

refinement or improvement of mind, tastes, etc., by education and training.
(Pocket Oxford Dictionary)

All organizations incorporate a set of cultures that have developed over a period of time. They are associated with the accepted standards of behaviour within the organization, and the development of a specific culture with regard to, for instance, quality, customer service and written communication, is a continuing quest for many organizations.

ESTABLISHING A SAFETY CULTURE – THE PRINCIPLES INVOLVED

With the greater emphasis on health and safety management implied in the MHSWR, attention should be paid by managers to the establishment and development of the correct safety culture within their organizations.

Both the HSE and the CBI have provided guidance on this issue. (J. R. Rimington, *The Onshore Safety Regime, The HSE Director General's Submission to the Piper Alpha Inquiry,* December 1989). The main principles involved, which involve the establishment of a safety culture, that are accepted and observed generally are:

● the acceptance of responsibility at and from the top, exercised through a clear chain of command, seen to be actual and felt throughout the organization

- a conviction that high standards are achievable as a result of proper management
- setting and monitoring of relevant objectives/targets, based on satisfactory internal information systems
- systematic identification and assessment of hazards and the devising and exercise of preventive systems that are subject to audit and review (in such approaches, particular attention is given to the investigation of error)
- immediate rectification of deficiencies
- promotion and reward of enthusiasm and good results.

DEVELOPING A SAFETY CULTURE – ESSENTIAL FEATURES

The following is an excerpt from *Developing a Safety Culture* by the CBI (1991).

> *Several features can be identified from the study that are essential to a sound safety culture. A company wishing to improve its performance will need to judge its existing practices against them.*
>
> 1 *Leadership and commitment from the top which is genuine and visible. This is the most important feature.*
> 2 *Acceptance that it is a long-term strategy which requires sustained effort and interest.*
> 3 *A policy statement of high expectations and conveying a sense of optimism about what is possible, supported by adequate codes of practice and safety standards.*
> 4 *Health and safety should be treated as other corporate aims, and properly resourced.*
> 5 *It must be a line management responsibility.*
> 6 *'Ownership' of health and safety must permeate at all levels of the workforce. This requires employee involvement, training and communication.*
> 7 *Realistic and achievable targets should be set and performance measured against them.*
> 8 *Incidents should be thoroughly investigated.*

9 *Consistency of behaviour against agreed standards should be achieved by auditing and good safety behaviour should be a condition of employment.*
10 *Deficiencies revealed by an investigation or audit should be remedied promptly.*
11 *Management must receive adequate and up-to-date information to be able to assess performance.*

PROMOTING A SAFETY CULTURE

The model given in Figure 10.1 shows the factors of cooperation, communication and competence within an overall framework of control.

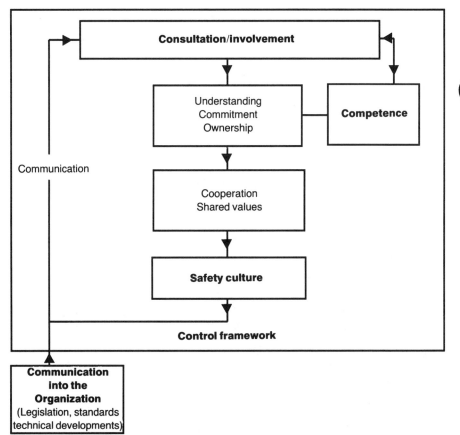

- **FIG 10.1 How to promote a positive health and safety culture – cooperation, communication and competence within a framework of control**

The process commences with a system for the communication of current legal requirements, standards and technical developments in the field of occupational health and safety within the organization. This communication system is an essential feature of the control framework.

The control framework is based on a process of consultation with, the involvement of, staff, contractors and others having a direct or indirect influence on health and safety procedures and systems. At this stage, the question of individual competence of people like health and safety practitioners, trade union safety representatives, managers and others will arise and may indicate a need for training of such people. The principal objective is to bring about a state of understanding, commitment and ownership of health and safety at all levels.

This state of ownership leads to cooperation on health and safety issues and shared values, which is the basis for a health and safety culture within the organization.

THE ROLE OF MANAGEMENT

As the legal requirements relating to health and safety at work move away from the concept of prescriptive standards to a more management and human factors-orientated approach, the role of senior management in developing and sustaining an appropriate safety culture becomes increasingly significant. What must managers do to encourage a positive safety culture?

First, the Board must clearly state its intentions, expectations and beliefs in relation to health and safety at work. In other words, it must state where they want the organization to be in terms of health and safety, and formulate action plans for achieving these objectives.

Adequate resources, in terms of financial resources, time and effort, must be made available in order to translate these plans and objectives into effective action. In particular, managers at all levels must be made accountable and responsible for their performance, as with other areas of performance, as part of this process. This should take place by means of routine performance monitoring and review, such performance being related to the reward structure of the organization. On-the-job performance monitoring should take into account the human decision-making components of a job, in particular the potential for human error.

Above all, senior managers and directors must be seen by all concerned to be taking an active and continuing interest in the development and implementation of health and safety improvements. On this basis, they should reward positive achievement in order to reinforce their message to subordinates that health and safety are of prime importance in the activities of the organization.

In the same way, the various lower levels of management must be actively involved. They must accept their responsibilities for maintaining health and safety standards as line managers and ensure that health and safety keeps a high profile within their area of responsibility. This will entail vigilance on their part to ensure, for instance, that safe systems of work are being followed, that people under their control are wearing the appropriate personal protective equipment and that unsafe practices by workers are not being adopted. They must show that deviations from recognized health and safety standards will not be tolerated, but, in doing so, it is important for line managers to recognize that they will receive backing from senior managers where such deviations actually occur. On this basis, it is vital that senior management demonstrate their commitment, too.

10

CONCLUSION

Establishing, developing and promoting the right safety culture is a prerequisite to good health and safety management and to ensuring compliance with the MHSWR.

The benefits of such a culture are manifest–reduced accidents, sickness absence, lost time and insurance premiums, increased overall performance, morale and commitment. This new caring culture should lead to greater profitability for the organization. It is a matter that cannot be overlooked.

Bibliography and further reading

Chapter 1

Health and Safety Executive, *Human Factors in Industrial Safety* (HS(G)48) (HMSO, 1989)

Chapter 2

McGregor, D., *The Human Side of Enterprise* (McGraw–Hill, New York, 1960)

Bass, B. M., *Organisational Psychology* (Allyn and Bacon, Boston, 1965)

Lippitt, R., and White, R. K, 'An experimental study of leadership and group life', in Eleanor E. Macoby, T. M. Newcomb and E. L. Hartley (eds), *Readings in Social Psychology* (Holt, New York, 1958)

Chapter 3

Health and Safety Executive, *Human Factors in Industrial Safety* (HS(G)48) (HMSO, 1989)

Health and Safety Commission, Advisory Committee on the Safety of Nuclear Installations, Second Report of Study Group, *Human Reliability Assessment – A critical overview* (HMSO, 1987)

Chapter 4

Katz, D., and Braly, K. 'Racial stereotypes of one hundred college students', *J. Abnorm. Soc. Psychol.*, 1933, 28, pp 280–290

Festinger, L., *A Theory of Cognitive Dissonance* (Harper & Row, New York, 1957)

Taylor, F. W., *Principles of Scientific Management* (Harper & Row, New York, 1911)

Mayo, E., *The Social Problems of an Industrial Civilisation* (Routledge & Kegan Paul, 1952)

Maslow, A. H., *Motivation and Personality* (Harper & Row, New York, 1954)

Herzberg, F., Mansner, B., and Snyderman, B. B., *The Motivation to Work* (John Wiley, New York, 1959)

Gregory, R. L., *Eye and Brain: The psychology of seeing* (McGraw-Hill, New York, 1966)

Allport, G. W., *Pattern and Growth in Personality* (Holt, Rinehart & Winston, New York, 1961)

Atkinson, J. W., and Feather, N. T., *A Theory of Achievement Motivation* (John Wiley, New York, 1966)

McClelland, D. C., Atkinson, J. W., Clark, R. A., and Lowell, E. L., *The Achievement Motive* (Appleton-Century-Crofts, New York, 1953)

Health and Safety Executive, *Human Factors in Industrial Safety* (HMSO, 1989)

Chapter 5

Heinrich, H. W., *Unsafe Acts and Conditions* (McGraw-Hill, New York, 1931)

Hale, A. R., and Hale, M., 'Accidents in Perspective' *Occupational Psychology*, 1970, 44, pp 115–121

Hale, A. R., and Hale, M., 'A review of the industrial accident research literature', Committee on Health and Safety at Work research paper (HMSO, 1972)

Powell, P. I., Hale, M., Martin, J., and Simon, M., *2000 Accidents* (National Institute of Industrial Psychology, 1971)

Chapter 6

Dempsey, P. J. R., *Psychology and the Manager* (Pan, 1973)

Warr, P. B., *Psychology at Work* (Penguin, 1971)

Swingle, P. G., *Social Psychology in Everyday Life* (Penguin, 1973)

Chapter 7

Health and Safety Commission *Management of Health and Safety at Work Regulations 1992 and Approved Code of Practice* (HMSO, 1992)

Department of Employment, *Safety Training Needs and Facilities in One Industry* (HMSO, 1973)

Department of Employment and Productivity, *Glossary of Training Terms* (HMSO, 1978)

Stranks, J. *Handbook of Health and Safety Practice* (third edition) (Pitman, 1994)

Chapter 8

Osborne, D. J., *Ergonomics at Work* (Wiley, 1982)

Sell, R. G., and Shipley, P., *Satisfaction in Work Design: Ergnomics and other approaches* (Taylor & Francis, 1979)

Health and Safety Executive, *Essentials of Health and Safety at Work* (HMSO, 1989)

Stranks, J., *Handbook of Health and Safety Practice* (third edition) (Pitman, 1994)

Grandjean, E., *Fitting the Task to the Man: An ergonomic approach* (Taylor & Francis, 1980)

Bell, C. R., *Men at Work* (Allen & Unwin, 1974)

Health and Safety Executive, *Workplace (Health, Safety and Welfare) Regulations 1992 and Approved Code of Practice* (HMSO, 1992)

Health and Safety Executive, *Lighting at Work* (HS(G)38) (HMSO, 1987)

Lyons, S., *Management Guide to Modern Industrial Lighting* (Butterworths, 1984)

Health and Safety Executive, *100 Practical Applications of Noise Reduction Methods* (HMSO, 1983)

Health and Safety Executive, *Noise Guides* (1–8) (HMSO, 1990)

Health and Safety Executive, *Manual Handling Operations Regulations 1992 and Guidance* (HMSO, 1992)

Pheasant, P., and Stubbs, D., *Lifting and Handling: An ergonomic approach* (The National Back Pain Association, 1992)

Health and Safety Commission, Health Service Advisory Committee, *Guidance on Manual Handling of Loads in the Health Service* (HMSO, 1992)

Health and Safety Executive, *Health and Safety (Display Screen Equipment) Regulations 1992 and Guidance* (HMSO, 1992)

Health and Safety Executive, *Visual Display Units* (HMSO, 1983)

Central Computer and Telecommunications Agency and the Council of Civil Service Unions, *Ergonomic Factors Associated with the Use of Visual Display Units* (CCTA, 1988)

National Electronics Council, *Human Factors and Information Technology* (National Electronics Council, 1983)

Chapter 9

Selye, Dr H., *The Stress of Life* (McGraw-Hill, New York, 1976)

Mackay, C., and Cox, T., *A Transactional Approach to Occupational Stress* (University of Nottingham, 1976)

Cox, T., *Stress* (Macmillan, 1978)

Barlow, J. M., *The Stress Manager* (Time Manager International (USA) Inc., San Francisco, 1986)

Spurgeon, A., *Stress at Work* (Industrial Safety Data File G: 16:1) (United Trade Press, 1987)

Stubbs, R., *Stress and Health at Work: Stress in British industry* (BUPA, 1987)

Coleman, V., 'Stress Management', *Training and Development*, 1922

Saville and Holdsworth, 'Stress Management', *Personnel Management*, 1988

Bond, M., and Kilty, J., *Practical Methods of Dealing with Stress* (University of Surrey 1982)

Allport, G. W., *Pattern and growth in personality* (Holt, Rinechart and Winston, 1961)